...HNIQUES IN MODERN GEOGRAPHY No. 46

GOODNESS-OF-FIT STATISTICS

by

A. S. Fotheringham and Daniel C. Knudsen

A. S. Fotheringham
Department of Geography
University of Florida
Gainesville, FL 32611
USA

Daniel C. Knudsen
Department of Geography
Indiana University
Bloomington, IN 47405
USA

British Library Cataloguing in Publication Data

Fotheringham, A. Stewart
Goodness-of-fit statistics in geographic research.——(Concept and technique in modern geography, ISSN 0306-6142).
1. Geography——Statistical methods
I. Title II. Knudsen, Daniel C. III.Series
910'.28 G70.3

ISBN 0-86094-222-8

1. INTRODUCTION

In any discipline, an important component of model building is the assessment of a model's ability to replicate a known data set. The accurate replication of a known data set by a model aids in validating the theoretical propositions on which the model is based. For instance, we may attempt to replicate migration patterns using data on distances and populations, or locational rents with data on distances from the centre of a city, or the geometry of stream channels with data on sediment loads. In each case it is important to be able to measure the accuracy of a model, since inaccuracy suggests we do not fully understand the process we are trying to model. Inaccurate replication of data also decreases one's confidence in using a model to predict unknown data, or to predict the effect of changes in a system. For instance, if a model can accurately replicate the revenue generated by a shopping centre based on its size, the model can be used to predict the additional revenue generated by adding, say, 10 000 square meters to the shopping centre. If the model is inaccurate, however, our confidence in predicting future revenue is diminished.

The failure of a model to replicate accurately an observed data set may be the result of one of several things: violations of the model's underlying assumptions, truly anomalous geographical behaviour, or evidence of model misspecification. Consequently, judging goodness-of-fit of a model is a complex task that is only partly a statistical problem. Statistics do, however, play a vital role in the assessment process. This monograph attempts to provide insights into the statistical aspects of this process.

The statistical assessment of how well a model replicates observed data involves use of one or more 'goodness-of-fit' statistics, that involve a quantitative description of some aspect of the difference between $\underline{\hat{Y}}$, the set of model predictions, and $\underline{Y}$, the set of observed values of the dependent variable in the model. At times, the choice of goodness-of-fit statistics can be critical. For example, Willmott (1984) employs three goodness-of-fit measures to evaluate model performance and each statistic indicates a different model as the most accurate. Because of a poor selection of goodness-of-fit statistic, Ayeni (1982) concludes that a model's estimates and the observed data are not significantly different at the 99% significance level, yet the model explains only 29% of the observed variance in the data! Contradictions such as these arise because different goodness-of-fit statistics measure different things, and hence they can yield different conclusions about the degree of correspondence between predicted and observed values of the dependent variable in a model. In this monograph we cannot discuss every goodness-of-fit statistic but we can examine the accuracy of, and the significance testing procedures associated with, a representative sample of such statistics.

Goodness-of-fit statistics serve two purposes. The first concerns comparison: either comparison of the accuracy with which two or more models replicate a known data set, or the comparison of the accuracy with which one model replicates two or more data sets. In either case, the relationship between error and the value of the goodness-of-fit statistic employed must be known in order to draw accurate conclusions regarding comparative model performance. A common mistake, for example, is to conclude that a model yielding a goodness-of-fit statistic twice as large as that produced by an alternative model is half (or twice, depending on the statistic) as accurate. As we shall demonstrate, such a conclusion is likely to be incorrect since the relationship between the value of a goodness-of-fit statistic and what we think of as error is often non-linear.

We will assume that the type of model comparison being undertaken is the typical situation in which nested hypotheses are being examined. That is, the models being compared can be considered as special cases of the same general model form. A common example is when the two models, A and B, are being compared, and model A contains all the variables in model B plus one or more additional variables. Another common example is when the same variable occurs in a different functional form in two otherwise identical models. For example, one model of shopping destination choice might postulate this choice as a power function of the cost of overcoming the spatial separation of an origin i and a destination j, ${c_{ij}}^{\beta}$, while another might postulate the same choice as an exponential function of cost, exp (βc_{ij}). Both models can be stated in the same general form as: 'choice = f (cost)', where f is some function to be specified. In many cases, it is a matter for empirical research and accurate model comparison to establish the most suitable functional form for a particular variable. The assumption of comparing nested hypotheses allows us to concentrate on goodness-of-fit statistics that are commonly employed in geographic research. A special set of goodness-of-fit statistics for the comparison of non-nested hypotheses has been developed and some of these are discussed by Anselin (1984). Such statistics are likely to see limited use in geography, however, where the comparison of truly non-nested hypotheses is very rare. As Anselin notes, even model forms that appear to be quite diverse can often be derived from the same general framework.

The second purpose served by goodness-of-fit statistics concerns hypothesis testing and the determination of whether the difference between sets of actual and predicted values is statistically significant. This demands knowledge of the sampling distribution of the statistic(s) used. Even when sampling distributions of the goodness-of-fit statistics are known, however, goodness-of-fit tests may produce undesirable results. For example, standard goodness-of-fit tests employing the chi-square statistic will reject the null hypothesis that the differences between $\underline{Y}$ and $\underline{\hat{Y}}$ are not significant (and hence we might conclude that the model is unsuitable) due to some trivial departure of the model estimates from the data if the sample size is sufficiently large. Alternatively, if the sample size is small, this same test may be unable to reject the null hypothesis, even when the departure of the model estimates from the observed data is quite large. This characteristic of some significance tests produces a number of practical problems in model evaluation (Openshaw, 1979).

Given these initial observations on the importance and the use of goodness-of-fit statistics, it might be expected that such a topic has been subjected to fairly intensive investigation, and that a standard set of goodness-of-fit measures has been established. Unfortunately, relatively few surveys or systematic examinations of goodness-of-fit statistics exist (examples include Smith and Hutchinson, 1981, Anselin, 1984, and Willmott, 1984), and the lack of investigation into the suitability of the numerous goodness-of-fit statistics available has led to a less-than-rigorous use of such statistics. There has, for example, been widespread use of *ad hoc* statistics with unknown sampling distributions, and use of statistics whose sensitivity to error is unknown; there has also been a lack of consistency in the use of different statistics in different studies (cf. Hathaway, 1975; Thomas, 1977; Fotheringham and Williams, 1983; Miller and O'Kelly, 1983; Southworth, 1983; Constanzo and Gale, 1984), which has hampered the comparison of results across studies.

In this monograph, particular emphasis is placed on identifying statistics that facilitate accurate comparison across data sets and/or models, and on identifying the conditions under which significance tests may produce undesirable results. While it may be impossible to identify one ideal goodness-of-fit statistic for all situations, we aim to provide information on the relative merits of particular statistics in a variety of circumstances. Hopefully, this will contribute to a greater awareness of the importance of selecting appropriate goodness-of-fit statistics in geographic research.

2. TWO GENERAL MODELS

In the subsequent section we review, and demonstrate the use of, several goodness-of-fit statistics commonly employed in geographic research to assess model validity. To facilitate an understanding of these statistics we will consider two general models, and an application of both.

2.1 An Additive Linear Model with Normally-Distributed Error

Probably the most common model form in geographic studies is an additive linear model, the simplest example of which is:

$$y_i = \alpha + \beta x_i + e_i, \tag{1}$$

where y_i is the dependent or endogenous variable, x_i is the independent or exogenous variable, α and β are parameters to be estimated given data on y_i and x_i, and e_i is a normally-distributed error term. It is usual to estimate the parameters of this model by least squares regression if the assumptions underlying this calibration method are met (see Ferguson, 1977). When estimates of α and β are obtained, estimates of y_i can be derived from:

$$\hat{y}_i = \hat{\alpha} + \hat{\beta} x_i, \tag{2}$$

where the symbol ^ denotes an estimated value. In this form, the model can be used to estimate data on y where it is unavailable, or to forecast changes in y due to changes in x.

In the survey of goodness-of-fit statistics that follows, we will employ this model form with data given in Table 1. The dependent variable, y, represents mean family income (mapped in Figure 1), and the independent variable, x, represents mean years of education. Data on both variables are given for 20 zones within a city.[1] A simple regression yields the following

[1]These data are taken from Willemain (1980, p.198), and are actual income and education data for 20 communities in Massachusetts. The spatial representation in Figure 1 is hypothetical.

Table 1: Zonal Data on Mean Income and Education

Zone (i)	Mean Family Income (y_i)	p_i	Mean Education (x_i)	Predicted Mean Income ($\hat{y}_i$)	q_i
1	10 599	0.044	11.4	10 215	0.043
2	10 621	0.044	11.8	11 218	0.047
3	9 507	0.040	10.4	7 707	0.032
4	17 558	0.073	13.5	15 482	0.064
5	11 415	0.048	12.3	12 472	0.052
6	15 609	0.065	12.8	13 726	0.057
7	11 631	0.048	12.4	12 723	0.053
8	9 691	0.040	10.2	7 205	0.030
9	10 277	0.043	11.4	10 215	0.043
10	9 861	0.041	12.0	11 720	0.049
11	12 412	0.052	12.4	12 723	0.053
12	7 146	0.030	12.0	11 720	0.049
13	10 677	0.045	12.4	12 723	0.053
14	12 382	0.052	12.1	11 971	0.050
15	23 430	0.098	15.2	19 746	0.082
16	17 361	0.072	14.9	18 993	0.079
17	8 973	0.037	11.1	9 463	0.039
18	9 494	0.040	10.4	7 707	0.032
19	10 203	0.043	11.3	9 964	0.042
20	11 091	0.046	12.2	12 221	0.051
Sample mean	11 997		12.1	11 997	
Sample standard deviation	3 799		1.3	3 272	

Source: Willemain (1980, p.198)

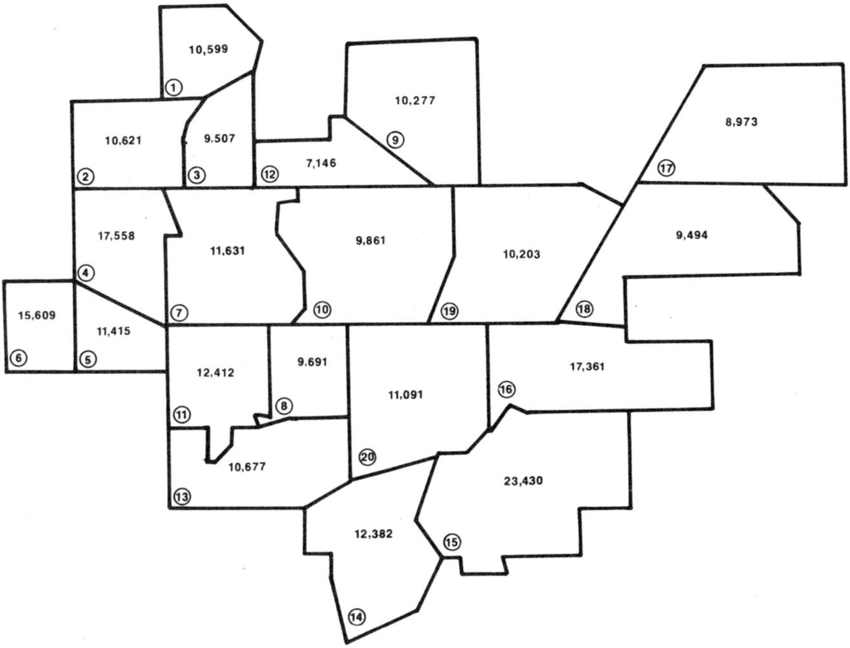

Figure 1. Mean Family Income by Zone

relationship between $\hat{y}_i$ and x_i:

$$\hat{y}_i = -18376 + 2508\ x_i,$$

with values of $\hat{y}_i$ given in Table 1. The task of the following section will be to identify statistics that accurately identify the degree of difference between $\underline{Y}$, the set of observed mean incomes, and $\hat{\underline{Y}}$, the set of predicted mean incomes. A plot of y against $\hat{y}$ is given in Figure 2. The latter figure is useful in showing the accuracy with which particular values of y_i are estimated by the model. Points close to the 45° line indicate values of y_i accurately estimated by the model; points far away from this line indicate values that are inaccurately estimated.

It is also useful to examine the spatial distribution of the regression residuals (Figure 3). If there is a significant clustering of either negative or positive residuals, this is often indicative of a poorly specified model in which an explanatory variable has been omitted, and which invalidates the results of the model calibration. In Section 6 we will describe a technique to examine the clustering of residuals more objectively, but for now we simply rely on the visual appearance of the residuals. There does not appear to be any strong clustering, and the negative and positive residuals, indicating under- and overpredictions respectively, seem fairly randomly distributed throughout the study area.

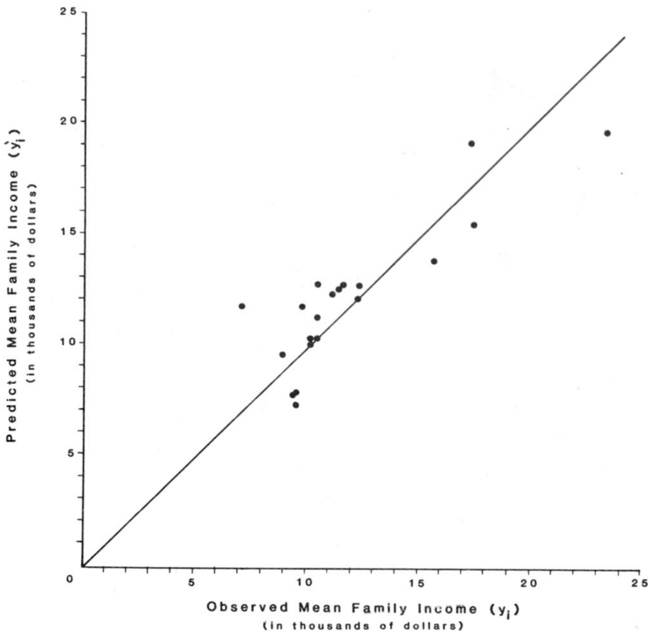

Figure 2. Observed Versus Predicted Mean Family Income

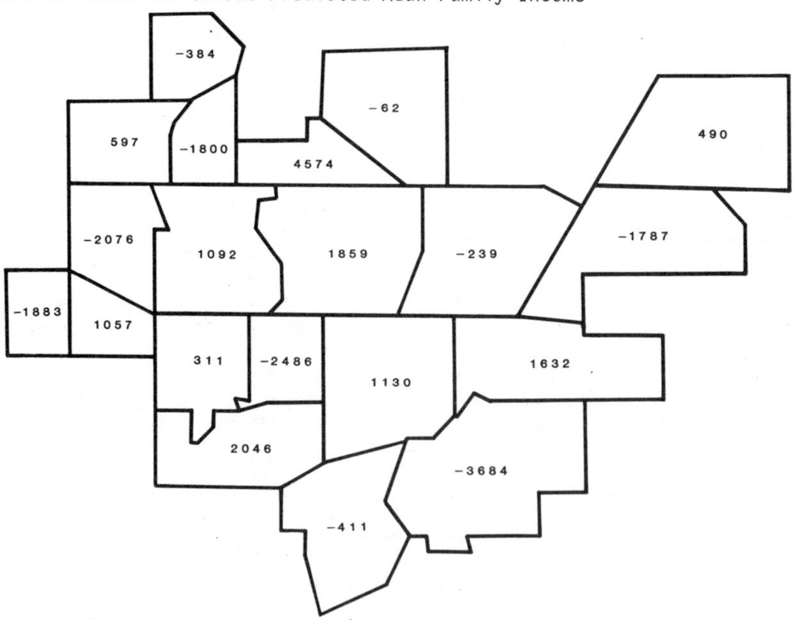

Figure 3. Mean Income Residuals ($\hat{y}_i$ - y_i) from regression

For some statistics that we consider, the observed and predicted values need to be converted to proportions, by dividing each value in a set by the sum of the values in that set. In this case, the sum of the observed values, T, is equal to the sum of the predicted values, $\hat{T}$, (a property of regression) and both totals are equal to 239 940. Thus, each observed and predicted value is divided by 239 940 to yield two new data sets, $\underline{P}$ and $\underline{Q}$, where $\underline{P}$ represents the set of observed proportions and $\underline{Q}$ represents the set of predicted proportions. The elements of $\underline{P}$ and $\underline{Q}$ are denoted by p_i and q_i, respectively and are given in Table 1 for the actual and predicted incomes.

2.2 A Multiplicative Model with Poisson-Distributed Error

Another common form of geographic model is the multiplicative form shown below, where the dependent variable is a count of some observations and the independent variable can either be count data or the more traditional continuous data:

$$y_i = \exp(\alpha)\, x_i^{\beta}\, e_i \,. \tag{3}$$

In geographic studies y_i could be, for example, a measure of the number of sinkholes in a limestone area, the number of work trips generated by a particular neighbourhood within a city, the number of trees of a particular type, the number of customers at a shop, and so on. In such cases, when the distribution of y_i is clearly not continuous, it is not appropriate to consider each error term being drawn from a normal distribution. Instead, we consider each y_i, which is a count of occurrences, to result from an unknown number of binomial experiments (the occurrence or non-occurrence of whatever is being measured by y_i). This leads to the assumption that each y_i, and hence each e_i, is the result of a random drawing from independent and identical Poisson distributions with mean and variance (λ) a common class of models that have the structure described in equation (3) are log-linear models (Upton and Fingleton, 1979; Wrigley, 1985). Another common model in geography, the gravity model of spatial interaction, can be constructed so that it has a Poisson-distributed error term (Flowerdew, 1982; Flowerdew and Aitkin, 1982; Fotheringham and Williams, 1984; Lovett, Whyte and Whyte, 1985).

To demonstrate the application of various goodness-of-fit statistics to models with Poisson error terms, we will use the following data which we imagine to be the number of sinkholes in various sections of a limestone region:

$$\underline{Y} = (10, 2, 7, 5, 9, 1, 5, 2, 4, 5) \,.$$

Imagine too that we have data on the density of the rock in each section and that the model described in equation (3) is fitted to the data, generating the following predicted values:

$$\underline{\hat{Y}} = (8, 5, 6, 4, 8, 2, 2, 3, 4, 8) \,.$$

Again, for some of the statistics discussed, we need to use proportions rather than the actual count data. In this case, the observed and predicted proportions are:

$\underline{P}$ = (.20, .04, .14, .10, .18, .02, .10, .04, .08, .10);

and,

$\underline{Q}$ = (.16, .10, .12, .08, .16, .04, .04, .06, .08, .16),

respectively.

3. GOODNESS-OF-FIT STATISTICS

A review of the geographic literature reveals that numerous goodness-of-fit statistics have been employed by geographers to assist in model validation. These statistics can be classified into four types: traditional; general distance; log-likelihood; and information-based. Representative statistics in each of these groups are now examined with reference to the two models discussed in the previous section.

3.1 Traditional Statistics

The two most commonly used goodness-of-fit statistics are the coefficient of determination (*inter alia* Hanushek and Jackson, 1977; Taylor, 1977; Silk, 1979; Clark and Ballard, 1980; Fotheringham, 1983; Jones, 1984), and the Pearson chi-square (*inter alia* Bishop *et al.*, 1975; Black and Salter, 1975; Hathaway, 1975; Openshaw, 1976; Pindyck and Rubinfeld, 1976; Snickars and Weibull, 1977; Baxter and Ewing, 1979). The former should only be employed for models with normally-distributed error; the latter only for models with Poisson-distributed error. We will thus demonstrate the use of the coefficient of determination with model 1, and the use of chi-square with model 2.

3.1.1 The Coefficient of Determination (R^2)

The expression for R^2 can be written in several different forms. Perhaps the most useful interpretative form is:

$$R^2 = \frac{\sum_{i=1}^{n} (y_i-\bar{y})^2 - \sum_{i=1}^{n} (y_i-\hat{y}_i)^2}{\sum_{i=1}^{n} (y_i-\bar{y})^2} , \qquad (4)$$

where $\bar{y}$ represents the mean of the ys, and n is the number of observations of y_i. In words, this expression is:

$$R^2 = \frac{\text{sum of squared residuals from the sample mean} - \text{sum of squared residuals from the regression line}}{\text{sum of squared residuals from the sample mean}} \qquad (5)$$

The statistic R^2 ranges between 0 and 1. If the regression line replicates each value of y_i perfectly, the sum of squared residuals from the regression line will be zero, and $R^2 = 1$. If the sum of squared residuals from the regression line is equal to the sum of squared residuals from the sample mean, then $R^2 = 0$. This will occur when the slope parameters of the regression model are equal to zero, and the estimated values of y_i are equal to the constant term which is $\bar{y}$. Thus, the R^2 statistic has two useful properties: it is bounded in both directions, and its limits have meaningful interpretations. R^2 has a third useful property, in that it can be interpreted as the proportion of the variance of y explained by the model. This interpretation can clearly be seen if the numerator and denominator of equation (4) are multiplied by 1/n. The equation can then be written as:

$$R^2 = \frac{\text{variance of y - variance of the residuals}}{\text{variance of y}} \quad . \qquad (6)$$

The lower is the variance of the residuals, the greater is the proportion of the variance of y explained by the model.

The formulation in equation (6), however, also indicates a potential weakness of R^2 in that the statistic is a function of the variance of y. This can lead to 'inflated' values of R^2 in a model where the observed data contain a few extremely large values. By 'inflated' we mean values of R^2 that are higher than we would expect given the size of the model residuals. These 'inflated' values of R^2 can be dangerous in giving the modeller unwarranted confidence in his/her model. We shall return to this point when we discuss error sensitivity in Section 4.

The calculation of R^2 for our income data discussed in Section 2 is presented in Table 2. The value of $R^2 = 0.74$ seemingly indicates a fairly high degree of model accuracy. That is, 74% of the variation in mean income can be explained by variation in education. However, we need to assess the probability that the relationship between $\underline{\hat{Y}}$ and $\underline{Y}$ could have occurred by chance. Actually, a variety of significance testing procedures is available to examine the statistical significance of R^2 under different circumstances. The first two tests we describe refer to the significance of the correlation coefficient (R), where $R = \sqrt{R^2}$, under the assumption that the values in $\underline{\hat{Y}}$ are derived from a simple linear regression model such as that given in equation (2). The third test assumes that the values in $\underline{\hat{Y}}$ are derived from a multiple linear regression model.

1. To examine the significance of the difference between the calculated value of R and zero, a t- test can be applied, where:

$$t_{n-2} = R/\sqrt{(1-R^2)/(n-2)} \quad , \qquad (7)$$

and where n-2 represents the degrees of freedom. For the income data, $R = \sqrt{.74} = \pm .86$, $n = 20$, and $t_{18} = \pm 7.156$. The two-tailed critical value of t at the 95% confidence level with 18 degrees of freedom is 2.101, and thus we can strongly reject the null hypothesis that the population correlation coefficient between $\underline{Y}$ and $\underline{\hat{Y}}$ is zero. That is, our model is producing a set of income predictions that is not completely random, but that exhibits some correspondence to the observed data. If we could not reject the null hypothesis that there was no

Table 2: Calculation of R^2 with Income Data

Zone	y_i	$\hat{y}_i$	$(y_i-\bar{y}_i)^2$	$(y_i-\hat{y}_i)^2$
1	10 599	10 215	1 954 130	147 303
2	10 621	11 218	1 893 100	356 887
3	9 507	7 707	6 199 600	3 239 280
4	17 558	15 482	30 925 800	4 309 780
5	11 415	12 472	338 608	1 118 100
6	15 609	13 726	13 047 300	3 544 180
7	11 631	12 723	133 833	1 192 900
8	9 691	7 206	5 317 180	6 177 220
9	10 277	10 215	2 958 060	3 819
10	9 861	11 720	4 562 070	3 455 880
11	12 412	12 723	172 308	96 845
12	7 146	11 720	23 531 200	20 921 500
13	10 677	12 723	1 742 140	4 186 930
14	12 382	11 971	148 302	169 085
15	23 430	19 746	130 716 000	13 574 800
16	17 361	18 993	28 773 600	2 664 070
17	8 973	9 463	9 143 970	239 905
18	9 494	7 707	6 246 510	3 192 660
19	10 203	9 964	3 218 080	56 930
20	11 091	12 222	820 656	1 278 260
			$\Sigma = 2.7186 \times 10^8$	$\Sigma = 6.99263 \times 10^7$

$$R^2 = \frac{2.7186 \times 10^8 - 6.99263 \times 10^7}{2.7186 \times 10^8} = 0.74$$

significant difference between R and zero, we might reject the theory that education is a determinant of income. However, this is a very insensitive test in terms of model validation. The set of predictions would have to be extremely poor in order not to reject the null hypothesis. In this case, where n = 20, rearranging equation (7) to solve for R, R would have to be less than 0.45 (and hence R^2 less than 0.2025) in order not to reject the null hypothesis of no significant relationship between $\hat{Y}$ and Y. Thus, while this significance testing procedure will identify extremely poor models, it will not adequately differentiate between the vast majority of models whose performance ranges from average to very accurate.

2. To examine the statistical significance of the difference between R and a value between 0 and 1, P, the test statistic is:

$$z = (Z_R - Z_P)/S(Z_R) \quad , \tag{8}$$

where z is a normal deviate, Z_R is Fisher's Z transformation:

$$Z_R = 1.1513[\log_{10}(1+R) - \log_{10}(1-R)] \quad , \tag{9}$$

with Z_P being defined in the same way and:

$$S(Z_R) = 1/\sqrt{n-3} \quad . \tag{10}$$

As Agresti and Agresti (1979) point out, the assumption that the test statistic in equation (9) has a Normal distribution is quite good when the sample size is at least 25 but is used in practice for sample sizes as small as 10. It would be very useful in a goodness-of-fit context to be able to examine the significance of the difference between R and 1.0 (a perfect model fit), but unfortunately this is not possible because $\log_{10}$ (1-R) is undefined when R = 1. However, the Fisher transformation described in equation (10) is useful to establish a confidence interval around Z_R and hence around R. The 100(1-α)% confidence interval for Z_ρ, where ρ is the population correlation coefficient between $\underline{Y}$ and $\underline{\hat{Y}}$, is:

$$Z_R \pm z_{\alpha/2}\ S(Z_R) \quad , \tag{11}$$

where z is the standard normal deviate corresponding to the $\alpha/2$ significance level in one-tail of the normal probability distribution. Once a confidence interval is established for Z_ρ, it is possible to establish a confidence interval for ρ by converting the end points of the Z_ρ confidence interval to values of ρ.

To demonstrate the use of this procedure in model validation, consider our income example. R = 0.86, and n = 20, therefore Z_R = 1.2934 and $S(Z_R)$ = 0.2425. Consequently, the 95% confidence interval for Z_ρ is 1.2934 ± 1.96 x 0.2425 = 1.2934 ± 0.4753 = 0.8181 to 1.7687. The values of R yielding these extremes of Z_R are 0.6740 and 0.9435 respectively, so that the 95% confidence interval for ρ is 0.6740 to 0.9435. Since the confidence interval contains only positive correlations, this reinforces the belief that income is positively related to education level. From the confidence interval of R, it is straightforward to obtain the confidence interval of R^2, the population coefficient of determination, by squaring the boundary points of the confidence interval for R. Thus, in our example, we can be 95% confident that R^2 lies between $(0.6740)^2$ and $(0.9435)^2$, that is, between 0.454 and 0.890. Hence, we can be 95% sure that there is at least a 45.4% reduction in error when we use education levels instead of the overall mean income figure to predict average incomes.

3. When the predicted values of our dependent variable are obtained from a regression model with more than one independent variable, we can examine the null hypothesis that R^2 = 0 with an F test, as follows:

$$F = \frac{R^2/k}{(1-R^2)\ /\ (n-k)} \quad , \tag{12}$$

where R^2 is the coefficient of multiple determination and k is the number of parameters (including the constant) estimated in the model. There are two degrees of freedom terms associated with the F test, df_1 and df_2, where df_1 = k and df_2 = n-k. We can again use this test to determine whether our calculated R^2 value is significantly different from zero, but it suffers from the same insensitivity problem as the t-test in that it can be used to identify poor models, but on its own cannot be used to classify a model as 'average' or 'good'. Models that yield surprisingly inaccurate predictions can 'pass' the F-test. However, the F-test can be used to compare the performance of different models.

While R^2 is thus a useful goodness-of-fit statistic and one that is very frequently encountered in geographic literature, it is not without its problems. Apart from the significance testing procedure problems identified above, there are several other potential problems associated with using R or R^2 as a goodness-of-fit statistic. Several authors have noted previously that R^2 is relatively insensitive to variations in model performance, and that R^2 can yield artificially high values when assessing model goodness-of-fit (Black and Salter, 1975; Wilson, 1976; McLafferty and Ghosh, 1982; Anselin, 1984; Willmott, 1984). Smith and Hutchinson (1981), for example, report values of R^2 as high as 0.70 even when $\underline{Y}$ and $\underline{\hat{Y}}$ differ by as much as 100%. Willmott (1984) demonstrates that the interpretations regarding model performance from the use of R^2 can be particularly misleading when the underlying relationship being modelled is not particularly linear. In such instances, the value of R^2 can be very high even when there are large differences between the observed and predicted data, or very low when there are only trivial differences.

R^2 is also an imperfect statistic to evaluate model performance across different data sets since, being a function of the variance of the observed data, its value is sample-specific. Also, it should be noted that R^2 measures the linear correlation between two data sets. Thus, if $\underline{Y}$ = (2, 3, 5, 1) and $\underline{\hat{Y}}$ = (4, 6, 10, 2), R^2 would obviously equal one although the predictions are not exact. This would not be a problem, however, if the predictions are generated from a model which incorporates a constraint on the total of the predicted values.

Frequently in geography, intrinsically linear models are calibrated by regression. A common example is when the model in equation (3) is made linear by taking logarithms. In such instances, a statistic that is useful to report with R^2 or R is $\hat{b}$, the estimated slope parameter, obtained from regressing $\underline{\hat{Y}}$ on $\underline{Y}$:

$$\hat{y}_i = a + by_i \quad . \tag{13}$$

In an accurate model, the estimated value of b, $\hat{b}$, should be close to 1.0 (see Figure 4). If $\hat{b}$ is less than 1.0, large values tend to be underpredicted by the model and small values overpredicted. If $\hat{b}$ is greater than 1.0, large values tend to be overpredicted by the model and small values underpredicted. Consequently, values of $\hat{b}$ that diverge from 1.0 indicate systematic errors in the modelling procedure. The divergence of $\hat{b}$ from 1.0 should be examined statistically with a t-test, where:

$$t_{\alpha/2,n-2} = \frac{\hat{b} - 1.0}{SE_{\hat{b}}} \quad , \tag{14}$$

and where:

$$SE_{\hat{b}} = \sqrt{\frac{\sum_i (y_i - \hat{y}_i)^2/(n-2)}{\sum_i (\hat{y}_i - \bar{y})^2}} \quad .$$

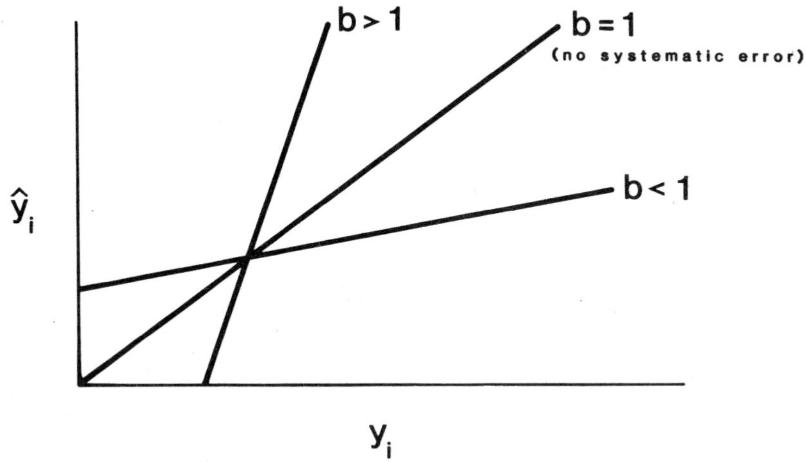

Figure 4. Systematic Errors and $\hat{b}$

When R^2 is being used to compare the performance of different models in a data set where the number of observations is small, the statistic should be adjusted to account for different degrees of freedom, these being related to the number of parameters estimated in a model. It is of little use, for example, to report that a model with 10 variables is more accurate than a model with 3 variables. To adjust for degrees of freedom, R^2 should be adjusted by:

$$R^2_{adj} = R^2 - \left(\frac{k-1}{n-k}\right) (1-R^2) \quad , \qquad (15)$$

where n represents the number of observations and k represents the number of parameters estimated in the model. The adjusted R^2 can be subjected to all the significance testing procedures described above. For our example:

$$R^2_{adj} = 0.74 - \left(\frac{2-1}{20-2}\right) (1-0.74)$$
$$= 0.74 - 0.014$$
$$= 0.726 \quad .$$

<u>3.1.2 Pearson Chi-Square</u> (χ^2)

For models with Poisson-distributed error terms, the Pearson chi-square statistic, χ^2, is defined as:

$$\chi^2 = \sum_{i=1}^{n} [(y_i-\hat{y}_i)^2/\hat{y}_i] \quad . \qquad (16)$$

The statistic has a lower limit of zero when $\hat{Y} = Y$ and an upper limit that tends towards positive infinity as any $\hat{y}_i$ tends towards zero.

For our sinkhole data, described in Section 2:

$$\begin{aligned}\chi^2 &= (10-8)^2/8 + (2-5)^2/5 + (7-6)^2/6 + (5-4)^2/4 \\ &+ (9-8)^2/8 + (1-2)^2/2 + (5-2)^2/2 + (2-3)^2/3 \\ &+ (4-4)^2/4 + (5-8)^2/8 \\ &= 9.3 \quad .\end{aligned}$$

Before discussing the significance of this particular result, it is important to note a critical difference between χ^2 and R^2. A perfect model fit is denoted by the minimum value of χ^2 but by the maximum value of R^2. Most goodness-of-fit statistics are similar to χ^2 in that increasing values indicate increasing error. This has important implications for the way in which significance tests for goodness-of-fit statistics are conducted. For statistics such as X^2 which represent perfect fit by a value of zero, it is standard practice in significance testing to set up a null hypothesis of no significant difference, and proceed to reject this null if the calculated value of the statistic exceeds the critical value at some predetermined significance level. This predetermined level is often 95% (α = 0.05) so that given a research hypothesis that significant differences actually exist, the chance of making a Type I error (rejecting the null hypothesis when it is true) is minimised. Such a confidence level, however, is poorly suited to the aim of assessing the accuracy of a model with statistics such as χ^2 since, in minimising Type I error, the procedure is extremely susceptible to Type II error. This latter error is particularly important in goodness-of-fit testing: we do not want to conclude falsely that our model is accurate. That is, we should make it as difficult as possible to achieve the result we want to achieve, which, in the context of goodness-of-fit testing, is usually that the model is an accurate one. By selecting a significance level of α = 0.05, we make it too easy to accept the null hypothesis that our model is accurate. Instead, when using a statistic such as χ^2, we should adopt a much more conservative significance level, one that reduces the risk of making a Type II error. Unfortunately, there is no standard level to minimize Type II error, as α = 0.05 has become the standard to minimize a Type I error. Here, we select α = 0.25 as the significance level whenever increasing values of the goodness-of-fit statistic indicate decreasing model accuracy. Thomas (1977), in an earlier CATMOG, makes this point regarding significance levels and goodness-of-fit testing, as do Abrahams and Mark (1986). The latter authors demonstrate that in model validation tests on channel networks, when the significance level was changed from α = 0.05 to the more appropriate α = 0.30, the proportion of the tests in which the model was rejected increased from 15 to 46 percent.

The degrees of freedom for the χ^2 test are n-k, where k is the number of parameters estimated in the model. For our data, the degrees of freedom are 8, and the critical value of χ^2 at α = 0.25 is approximately 10.25. Given our calculated value of χ^2 = 9.3, we thus accept the null hypothesis that there is no significant difference between $\underline{\hat{Y}}$ and $\underline{Y}$ and conclude that our model is an accurate one.

While the χ^2 test, when used with the appropriate significance level, is a useful goodness-of-fit statistic, there are two potential problems with its use:

1. Due to the division of the statistic by $\hat{y}_i$, it is sensitive to low predicted values of y. For this reason, some researchers adopt a general guideline that χ^2 should not be used when any value of $\hat{y}_i$ is less than five. Others use a less restrictive guideline that not more than 20% of the $\hat{y}_i$ values should be less than five and none should be less than one.

2. Because the units of the numerator are the square of the units in the denominator, the statistic is not standardised for different data magnitudes. For instance, if all the values in both $\hat{\underline{Y}}$ and $\underline{Y}$ are doubled, the value of χ^2 is also doubled. This would be acceptable if the degrees of freedom increased proportionately, but these are defined only in terms of the number of observed and predicted values and the number of parameters in the model, not in terms of the magnitude of the data. This problem with χ^2 is especially severe when the units of the observed values are subjectively defined. Such a situation arises in many geogrpahical studies: for example, one might predict, say, expenditure flows where the units of currency could be pounds, hundreds of pounds, dollars, etc., or agricultural yields which could be expressed as pounds, kilograms or tons, or sediment loads of streams which could be defined in terms of ounces or grams.

3.2 General Distance Statistics

We define general distance statistics as being measures of the average difference between y_i and $\hat{y}_i$. General distance statistics include Gini coefficients (Chu, 1982; Gaile, 1983), the Index of Dissimilarity (Duncan and Duncan, 1955; Timms, 1965; Thomas, 1977) and measures of root mean square error (Black, 1973; Openshaw and Connolly, 1977; Pitfield, 1978; Willmott, 1984). As representative of general distance statistics, we examine standardised root mean square error (SRMSE), which is defined by Pitfield (1978) as:

$$\text{SRMSE} = \frac{1}{\bar{y}} [\Sigma(y_i-\hat{y}_i)^2/n]^{1/2} \quad . \tag{17}$$

The statistic is the Root Mean Square Error standardised by dividing by $\bar{y}$. It has a lower limit of zero (indicating perfectly accurate predictions) and an upper limit that is variable and depends on the distribution of y_i, although in practice it is usually 1.0. Values of SRMSE greater than 1.0 only occur when the average error is greater than the mean. For example, if $\underline{Y}$ = (0, 0, 0, 2) and $\hat{\underline{Y}}$ = (2, 0, 0, 0), SRMSE = 2.828. For our sinkhole data:

y_i	$\hat{y}_i$	$(y_i-\hat{y}_i)^2$
10	8	4
2	5	9
7	6	1
5	4	1
9	8	1
1	2	1
5	2	9
2	3	1
4	4	0
5	8	9
50	50	36

$$\text{SRMSE} = \frac{1}{5}[36/10]^{1/2} = 0.379 \quad .$$

A major problem with the use of SRMSE (this also applies to the other general distance goodness-of-fit statistics) is that there is no theoretical distribution for the statistic, and hence it is difficult to interpret our value of 0.379 except to say it is closer to 0.0 than 1.0. We cannot assess the accuracy of our model further at this stage. However, in a later section we will discuss the derivation of experimental distributions which can be used to assess the significance of any statistic.

SRMSE is preferred to root mean square error, which is not divided by $\bar{y}$ because the latter is very sensitive to the magnitude of the data, and is therefore not comparable between spatial systems. The unstandardised statistic is also sensitive to large deviations from the mean (Wilson, 1976; Southworth, 1977; Pitfield, 1978). SRMSE should only be used when $\sum_i \hat{y}_i = \sum_i y_i$.

3.3 Information-Based Statistics

Suppose that we have two probability distributions, P and Q, and that initially we use Q to estimate P. Then, the information that we gain from actually knowing P as opposed to estimating it from Q is measured by Kullback and Leibler's (1951) information statistic:

$$I(P:Q) = \sum_i p_i \ \ln(p_i/q_i) \quad . \tag{18}$$

Information-based goodness-of-fit statistics are based on this equation because it is also a measure of the similarity between two sets of probabilities or proportions. Here, we will examine three such information statistics, Information Gain, Psi, and Absolute Entropy Difference. For more examples of the use of information statistics in geographic research, see Thomas (1981) and Johnston and Semple (1983).

3.3.1 Information Gain

The formula for Information Gain, I(P:Q), is that of equation (18), where p_i is known as a posterior probability (akin to the dependent variable in regression), and q_i is known as the prior probability (akin to the independent variable in regression). The statistic has a minimum of zero when P = Q, and a maximum at positive infinity whenever $p_i > 0$ and $q_i = 0$ for any i, j pair (see also Phillips, 1981: Thomas, 1981). In many instances when the elements of Q are model predictions, they cannot be zero and so the maximum value of I in such cases will be some large but finite number. In calculating I for our sinkhole data sets, converted into proportions, we take P to be the set of observed proportions and Q to be the set of predicted proportions:

$$\begin{aligned} I = {} & .2 \ln(.2/.16) + .04 \ln(.04/.1) + .14 \ln(.14/.12) \\ & + .1 \ln(.1/.08) + .18 \ln(.18/.16) + .02 \ln(.02/.04) \\ & + .1 \ln(.1/.04) + .04 \ln(.04/.06) + .08 \ln(.08/.08) \\ & + .1 \ln(.1/.16) \end{aligned}$$

$= .04463 - .03665 + .02158$

$+ .02231 + .02120 - .01386$

$+ .09163 - .01622 + 0 - .04700$

$= .08762$.

It is easy to see that underpredictions ($p_i > q_i$) yield positive values of $p_i \ln(p_i/q_i)$ and overpredictions ($p_i < q_i$) yield negative values. When $p_i = q_i$, $\ln(p_i/q_i) = 0$. Information gain is useful in emphasising errors in large values because each expression of error, $\ln(p_i/q_i)$, is weighted by a function of the actual value p_i.

The significance of information gain can be found through its relationship to the minimum discrimination information statistic (MDI):

$$MDI = 2T \times I(P:Q) \quad , \tag{19}$$

where:

$$T = \sum_i y_i \quad . \tag{20}$$

MDI is asymptotically chi-square distributed (Bishop *et al.*, 1975; Phillips, 1981) with (n-k) degrees of freedom, where k is the number of parameters in the model. In our case, MDI = 2 x 50 x 0.08762 = 8.762. The critical value of χ^2 (df = 8; α = 0.25) is approximately 10.25, so that at this level of confidence we would accept the null hypothesis of no significant difference between the predicted and the observed sinkhole occurrences. That is, the model from which the values of Q are derived appears to be a reasonable one for this particular data set.

Although information gain is widely used as a goodness-of-fit statistic and has some useful properties, there are two potential problems with its use. Firstly, the value of the statistic depends on which vector is defined as P and which as Q, since $I(P:Q) \neq I(Q:P)$ unless P = Q (Tribus and Rossi, 1973; Bishop *et al.*, 1975; Knudsen, 1982). In this instance, for example, I(Q:P) = 0.1577 and MDI = 15.77, so that we would reject the null hypothesis instead of accepting it. Secondly, when $p_i > 0$ and $q_i = 0$ for corresponding elements of P and Q, infinite values of information gain are produced. Neither of these problems, however, should arise in modelling, since P is defined as a vector of posterior (known) information and Q is defined as a vector of prior (predicted) information, which usually leaves little doubt about which is which. Also, as already mentioned, all of the predicted values may be non-zero. However, in instances where one or more of these problems arise, a superior goodness-of-fit statistic is available in the psi statistic described below.

3.3.2 Psi

The psi statistic was developed by Kullback (1959) and introduced into the geographical literature by Ayeni (1982; 1983). It is defined as:

$$\Psi = \sum_i p_i \ln(p_i/s_i) + \sum_i q_i \ln(q_i/s_i) \quad , \tag{21}$$

where $s_i = (p_i + q_i)/2$. The psi statistic has a lower limit of zero when $\underline{P} = \underline{Q}$, and an upper limit of (n ln 2) whenever the non-zero elements of $\underline{P}$ correspond to the zero elements of $\underline{Q}$, and *vice versa*. Unlike information gain, psi is insensitive to the designation of $\underline{P}$ and $\underline{Q}$ since it measures differences in $\underline{P}$ and $\underline{Q}$ with reference to $\underline{S}$, and it allows values of p_i and q_i to be zero.

The significance of psi can also be found from its relationship to the MDI statistic which is identical to that given in equation (8), that is, MDI = 2T x Ψ.

In calculating Ψ with our sinkhole data:

$$\underline{S} = (.18, .07, .13, .09, .17, .03, .07, .05, .08, .13),$$

and:

$$\begin{aligned}\Psi = {} & [.2 \ln(.2/.18) + .04 \ln(.04/.07) + 14 \ln(.14/.13) + \\ & .1 \ln(.1/.09) + .18 \ln(.18/.17) + .02 \ln(.02/.03) + \\ & .1 \ln(.1/.07) + .04 \ln(.04/.05) + .08 \ln(.08/.08) + \\ & .1 \ln(.1/.13)] + [.16 \ln(.16/.18) + .1 \ln(.1/.07) + \\ & .12 \ln(.12/.13) + .08 \ln(.08/.09) + .16 \ln(.16/.17) + \\ & .04 \ln(.04/.03) + .04 \ln(.04/.07) + .06 \ln(.06/.05) + \\ & .08 \ln(.08/.08) + .16 \ln(.16/.13)] \\ = {} & [.02107 - .02238 + .01038 + .01054 + .01029 - .00811 + \\ & .03567 - .00893 + 0 - .02624] + [-.01885 + .03567 - \\ & .00961 - .00942 - .00970 + .01151 - .02238 + \\ & .01094 + 0 + .03322] \\ = {} & .02229 + .02138 = .04367 \quad .\end{aligned}$$

Consequently:

$$\text{MDI} = 2 \times 50 \times 0.04367 = 4.367 \quad .$$

With the critical value of χ^2 at $\alpha = 0.25$ and 8 degrees of freedom being 10.25, we again accept the null hypothesis that there is no significant difference between the predicted and the observed sinkhole occurrences, and conclude that our model is an accurate one. The use of Ψ is more likely to lead to the acceptance of the null hypothesis that there is no significant difference between $\underline{P}$ and $\underline{Q}$ and, consequently, we are more likely to accept as valid the theory on which the model is based. While the use of I is more rigorous than the use of Ψ in testing theory, Ψ has the capacity to account for zero entries in the original data or in the set of predicted values.

There are also situations where neither the use of I nor Ψ is particularly informative in testing hypotheses. The MDI formulation used in testing both I and Ψ is a function of T, the total number of observations. As with the χ^2 test, if this value is subjectively defined, then the significance testing procedure using MDI is virtually meaningless. In situations where MDI is not particularly useful, the significance testing of I and Ψ can still be undertaken by deriving experimental distributions for the statistics using Monte Carlo methods, as will be discussed in Section 4.

3.3.3 Absolute Entropy Difference

A third information-based statistic is the absolute entropy difference (AED), which is defined as the absolute value of the difference in the entropies of the observed and predicted probability distributions:

$$AED = |H_P - H_Q| \quad , \tag{22}$$

where H denotes Shannon's entropy measure (1948), so that:

$$H_P = -\sum_i p_i \ln p_i \quad , \tag{23}$$

$$H_Q = -\sum_i q_i \ln q_i \quad . \tag{24}$$

The logic of the entropy differnece test is based on entropy being a measure of the variance of the data in P and Q. When P and Q are very similar, they will have similar variances and hence AED will be low. When P and Q are dissimilar they are likely (although see below) to have different variances and AED will be large. The minimum and maximum value of each of the entropy functions is zero and ln n, respectively, where n is the number of elements in P. Consequently, AED ranges between zero and ln n (Thomas, 1981). For our sinkhole data, $H_P = 2.134$ and $H_Q = 2.1948$ so that AED = 0.0604.

The statistical significance of AED can be examined if we assume that the entropy values are the means of normal distributions. A t test of differences between means can then be applied (Hutcheson, 1970). In the case of H_P, the entropy of the observed sinkhole occurrances, we are assuming that repeated measurements of these occurrences would yield an approximately normal distribution of H_P values. In the case of H_Q, the predicted sinkhole occurrences, we are assuming that repeated measurements of the observed would yield different parameter estimates in the model from which the predicted sinkhole occurrences are derived, and which in turn would yield an approximately normal distribution of the H_Q values. The test is:

$$t = \frac{|H_P - H_Q|}{\sqrt{[Var(H_P) + Var(H_Q)]/2}} \quad , \tag{25}$$

where:

$$Var(H_P) = \sum_i p_i(\ln p_i)^2 - H_P{}^2/T \quad , \tag{26}$$

and:

$$Var(H_Q) = \sum_i q_i(\ln q_i)^2 - H_Q{}^2/T \quad , \tag{27}$$

where T is the sum of the observations in Y (50 in our case). For our data:

$$\sum_i p_i(\ln p_i)^2 = 4.8244$$

$$\sum_i q_i(\ln q_i)^2 = 5.0062 \quad ,$$

so that:

$$Var(H_P) = .00537 \quad ,$$

and:

$$Var(H_Q) = .00378 \quad .$$

Consequently:

$$t = 0.0604/0.0676 = 0.8935 \quad .$$

At the 75% confidence level and with n-k degrees of freedom (8 in this case), the critical value of t 8s 0.706 and therefore we reject the null hypothesis that there is no significant difference between the observed and predicted values. We conclude that the model used to generate the predicted number of sinkholes is not a reasonable one for this data set.

Because AED is a function of the difference between two summary statistics, it can give misleadingly low values when $\underline{P}$ and $\underline{Q}$ are dissimilar. For example, if $\underline{P} = (0.2, 0.6, 0.1, 0.1)$ and $\underline{Q} = (0.1, 0.1, 0.6, 0.2)$, $H_P = H_Q$ despite the fact that the corresponding data elements in the two data sets are very dissimilar. Users of this statistic should thus be very careful to avoid the false conclusions that can arise from the use of AED. Other potential problems with the use of AED significance testing include the assumption that the entropy values are normally-distributed, and doubts have been raised regarding the validity of the variance formulas in equations (26) and (27) under certain circumstances (Pielou, 1969). Despite these shortcomings, AED can be a useful statistic because it appears to be a rigorous test of model validity, and because of its linear relationship with error (see Section 4 below).

3.4 Log-Likelihood Statistics

A final category of goodness-of-fit statistics involves the use of maximum likelihood as a general method of estimating the parameters of a model, in which parameters are found that are 'most likely' given the observed data (McCullagh and Nelder, 1983). For a general outline of likelihood and maximum likelihood, see Vincent and Haworth (1984) and Pickles (1986). To illustrate briefly the notion of likelihood, let us examine a classic example: that of drawing, with replacement, balls from an urn containing only pink and blue balls. Suppose that in five draws we obtain four pink balls and one blue ball. What is the most likely proportion of pink balls in the urn? Let the probability of drawing a pink ball on any one trial be δ and the probability of drawing a blue ball on any one trial be $1-\delta$. Because we are sampling with replacement, the drawing of each coloured ball is an independent event, and therefore the probability of observing 4 pink balls and 1 blue ball is $\delta^4(1-\delta)$. Following the logic of maximum likelihood estimation, we seek a value for the parameter δ that is most likely to return four pink balls and one blue ball in five draws. That is, we want to find $\hat{\delta}$ such that:

$$L = \hat{\delta}^4 \, (1-\hat{\delta})$$

is at a maximum. However, use of the likelihood function itself is cumbersome. Since the maximum of L and ln L coincide, it is convenient to maximise:

$$L^* = \ln L = 4 \ln \hat{\delta} + \ln(1-\hat{\delta}) \quad .$$

When L^* is at a maximum, the derivative of L^* with respect to $\hat{\delta}$, $\partial L^*/\partial\hat{\delta}$, is equal to zero. That is:

$$\partial L^*/\partial\hat{\delta} = 4/\hat{\delta} - 1/1-\hat{\delta} = 0 \quad ,$$

which, on rearranging, yields:

$$\hat{\delta} = 0.8 \quad .$$

For both the additive and multiplicative models that we have been using in this monograph, obtaining the maximum likelihood estimates for the parameters α and β proceeds in a similar fashion. When parameter estimates are obtained via maximum likelihood estimation, likelihood goodness-of-fit statistics follow naturally as means of model validation.

Typically, we evaluate a given model's performance relative to the performance of a 'full' model: a model containing one parameter for each observation of the dependent variable, y. As a result, this full model exactly replicates the original data. We may then take the ratio of a model's likelihood to that of the full model, or alternatively, take the difference of the logs of these likelihoods as a measure of model goodness-of-fit. A useful statistic in this regard is:

$$D = 2\ |L^*(1) - L^*(2)| \quad , \tag{28}$$

where $L^*(1)$ is the log-likelihood of the full model, and $L^*(2)$ is the log-likelihood of the model that generated $\hat{Y}$. Such a statistic has elsewhere been termed deviance (Nelder and Wedderburn, 1972; McCullagh and Nelder, 1983). To proceed further, we must evaluate the terms $L^*(1)$ and $L^*(2)$. This entails specifying the distribution of $\hat{y}$ and y. In our income model, where we have assumed the error terms to be Normally-distributed, $\hat{y}$ and y are Normal since linear functions of Normally-distributed variables are also Normally-distributed. When $\hat{y}$ and y have Normal distributions:

$$2L^*(1) = n\ \ln(2\pi\sigma_y^{\ 2}) + \sum_i (y_i-\tilde{y}_i)^2/\sigma_y^{\ 2} \quad , \tag{29}$$

and:

$$2L^*(2) = n\ \ln(2\pi\sigma_y^{\ 2}) + \sum_i (y_i-\hat{y}_i)^2/\sigma_y^{\ 2} \quad , \tag{30}$$

where $\tilde{y}_i$ denotes an estimate of y_i generated by the full model (model 1) and σ^2 represents a variance. First, we observe that by definition $\tilde{y}_i = y_i$ for all i, thus the right-most term of equation (29) is identically zero. Substituting (29) and (30) into equation (28) yields:

$$D = |n\ \ln(2\pi\sigma_y^2) - n\ \ln(2\pi\sigma_y^2) - \sum_i (y_i-\hat{y}_i)^2/\sigma_y^2\,| \tag{31}$$

$$= \sum_i (y_i-\hat{y}_i)^2/\sigma_y^2 \quad ,$$

which is the deviance term for models with Normal error terms. Notice that if the variables are normalised (z-scores), then $\sigma_y^2 = 1$ and the deviance reduces to a sum of squared errors. D is distributed as chi-square with degrees of freedom n-k, where k is the number of parameters in the model generating $\hat{Y}$. D has a minimum of zero when $\hat{\underline{Y}} = \underline{Y}$ and a maximum value that tends toward positive infinity.

For our income data, we know from Table 2 that $\sum_i (y_i-\hat{y}_i)^2 = 69\ 926\ 300$ and that $\sum_i (y_i-\bar{y})^2 = 271\ 860\ 000$. Since n=20:

$$\sigma^2_y = \sum_i (y_i-\bar{y})^2/20 = 13\ 593\ 000 \quad ,$$

and:

$$D = 69\ 926\ 300/13\ 593\ 000 = 5.144 \quad .$$

At the 75% confidence level and with 8 degrees of freedom, the critical value of the chi-square distribution is 10.25, so that we can accept the null hypothesis that $\hat{Y}$ and Y are not significantly different and that our model is an accurate one.

In our sinkhole example, the errors are not Normally-distributed and instead have Poisson distributions. This leads to different formulations of L*(1) and L*(2), which for the Poisson case are:

$$L^*(1) = \sum_i (y_i \ln \tilde{y}_i - \tilde{y}_i) \quad , \tag{32}$$

and:

$$L^*(2) = \sum_i (y_i \ln \hat{y}_i - \hat{y}_i) \quad . \tag{33}$$

Thus:

$$2\ |L^*(1) - L^*(2)| = 2\ |\sum_i y_i \ln \tilde{y}_i - \sum_i y_i \ln \hat{y}_i - \sum_i \tilde{y}_i + \sum_i \hat{y}_i \ | \quad .$$

Since $\tilde{y}_i = y_i$ and $\sum_i \hat{y}_i = \sum y_i$,

$$D = 2\ |\sum_i y_i \ln y_i - \sum_i y_i \ln \hat{y}_i\ |$$

$$= 2\ |\ y_i \ln(y_i/\hat{y}_i)\ | \quad , \tag{34}$$

which is the same formulation as the MDI statistic described in the section on information gain statistics. Consequently, D for the sinkhole data is 8.762, which at the 75% confidence level is less than the critical value of chi-square with 8 degrees of freedom. We thus accept the null hypothesis that $\hat{Y}$ and Y are not significantly different and conclude that our model is an accurate one.

Notice that the formulation for D given in equation (28) is a very general one that can be used to compare the performance of any two models, and not simply the performance of one model against a perfect model (that is, one that replicates the observed data exactly). It is thus a useful statistic for comparing the relevance of particular variables in a model: a comparison can be made between models where one or more variables have been excluded. The degrees of freedom in such an application would be equal to the difference in the number of parameters estimated in the two models.

4. ERROR SENSITIVITY OF GOODNESS-OF-FIT STATISTICS

We mentioned initially that goodness-of-fit statistics are useful in two situations: one is when we want to evaluate the performance of one model in replicating a data set; the other is when we want to evaluate either the performance of one model in replicating two or more data sets, or the performance of two or more models in replicating a single data set. For the former, we often rely on the significance testing procedure which we have discussed and to which we will return in the next section. For the latter, we often rely on an assumption that the relationship between the values of a goodness-of-fit statistic and error levels is linear. For instance, we might assume that an R^2 value of 0.9 indicates twice as accurate a model fit as an R^2 value of 0.45.

Knudsen and Fotheringham (1986) have recently described the error sensitivity of various goodness-of-fit statistics, and here we expand their results. An idea of the sensitivity of each statistic identified in Section 3 to variations in error levels is obtained through simulation, first under an assumption of Normal, additive errors and then under an assumption of Poisson-distributed, multiplicative errors. For Normal errors, this involves the generation of a vector of Normally-distributed data. Model estimates are then obtained at error levels of 1%, 5% and 10-100% in increments of 10%, using the transformation:

$$\hat{y}_i = y_i + \delta(\bar{y} \;.\; \text{RND} \;.\; \text{FACT}) \, , \qquad (35)$$

where $\hat{y}_i$ is the model estimate, y_i is the observed value, δ randomly takes on the value of $\pm$ 1, $\bar{y}$ is the mean of $\underline{Y}$, RND is a random number between zero and one, and FACT is the percentage error introduced divided by 100. By this construction, however, negative or zero values of y_i are likely, and these need to be replaced with small, positive values when calculating those statistics, such as information gain, where nonpositive values cannot be used. The subsequent results are generated with replacement values of unity.

For Poisson-distributed, multiplicative error, $\hat{y}_i$ values are generated by random drawing from a Poisson distribution, and model estimates are again generated at various error levels, in this instance utilising:

$$\hat{y}_i = y_i + \delta(y_i \;.\; \text{RND} \;.\; \text{FACT}) \, , \qquad (36)$$

where terms are as previously defined (cf. Smith and Hutchinson, 1979). In this instance, zero values of $\hat{y}_i$ are unlikely since they can only occur when δ = -1, RND = 1, and FACT = 100.

Results of the sensitivity analysis are given in Figures 5 and 6. The vertical axis on each figure represents the values of statistics standardised between 0 and 1. These standardised values are obtained by dividing each value by the largest value of the statistic obtained over a given error range. The horizontal axis represents the proportion of error separating $\hat{\underline{Y}}$ and $\underline{Y}$. Figure 5 represents the situation where error is additive and Normally-distributed; Figure 6 represents the situation where error is multiplicative and Poisson-distributed. With the former errors, the chi-square test is inappropriate and so is omitted: with the latter, R^2 is inappropriate

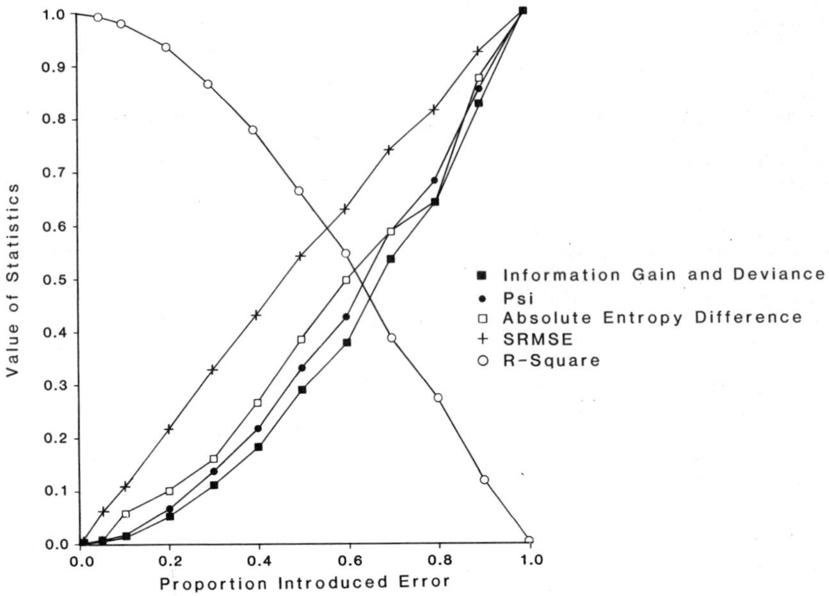

Figure 5. Error Sensitivity for Normal Error

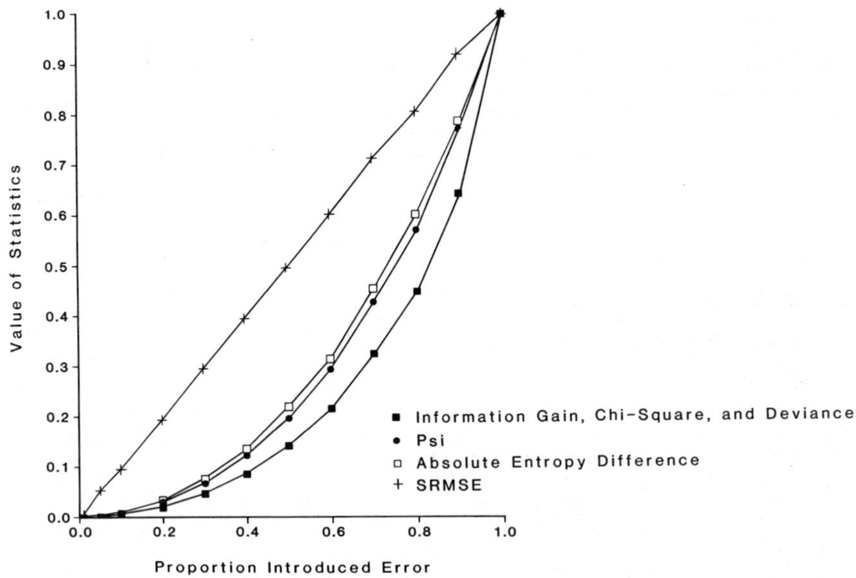

Figure 6. Error Sensitivity for Poisson Error

and is omitted. On either graph, a useful goodness-of-fit statistic for comparative purposes would be indicated by a linear relationship between the value of the statistic and error level. This property would imply that the statistic is equally responsive to all error levels, and that metric properties could be applied to the statistic. The relationships in Figure 5 indicate that the ranking of the statistics in order of their usefulness for comparative purposes for Normal error is: SRMSE, R^2, AED, psi, and information gain and log-likelihood difference. The relationships in Figure 6 indicate that the ranking of the statistics for Poisson error is: SRMSE, AED, psi, log-likelihood difference, information gain, and chi-square. This ranking suggests that generalised distance statistics such as SRMSE are superior to other types of goodness-of-fit statistic for comparative purposes (comparing the performance of the same model in different systems). This is not unexpected, since the error levels on the graph are defined in terms of percentage error. However, what may be surprising is the extent of the divergence of several of the statistics from a linear relationship. In the Normal error case, for example, the R^2 statistic has a value that is logarithmically related to error levels. As error levels increase, the value of R^2 decreases slowly at low error levels, but above an approximately 20% error level, the value of the statistic decreases rapidly and almost linearly. The lower bound of the statistic also appears to be conditional on the distribution of the observed flows (cf. Wilson, 1976; Silk, 1979; McLafferty and Ghosh, 1982; Knudsen and Fotheringham, 1983).

Chi-square, information gain, psi and AED also exhibit nonlinear relationships with error levels. This nonlinearity is consistently more pronounced for information gain, deviance and chi-square than for the remaining statistics, and hence the assumption of a linear relationship for these statistics could lead to misleading conclusions regarding the extent of the superiority of one model over another. Psi and AED lie between SRMSE and chi-square in therms of error sensitivity. Conclusions about relative model performance with these statistics would be more unreliable for Poisson error than for Normal error, and psi is more nonlinear than AED for both error distributions. Conclusions about model performance based on D may also be misleading, since the statistic is chi-square distributed for Normal error and only asymptotically chi-square for Poisson error.

5. SIGNIFICANCE TESTING AND GOODNESS-OF-FIT STATISTICS

5.1 Theoretical Significance Tests

Theoretical or 'standard' significance testing procedures exist for many, but not all, of the statistics that have been discussed. Such procedures make reference to a theoretical distribution. For example, the significance of the information gain statistic can be examined by calculating the value of the associated MDI statistic which is assumed to be chi-square distributed (in fact it is asymptotically chi-square distributed). Similarly, the significance of a particular R^2 value can be determined by reference to a Normal distribution. It is important to recognize two potential problems with this type of significance testing procedure.

(i) In identifying useful significance testing procedures for model evaluation, it is important to consider possible complications introduced by alternative definitions of y_i. For example, the same set of commodity flows may be measured in kilograms, tons, or rail carloads; the same set of shopping expenditures may be measured in dollars or hundreds of dollars; the same set of migration flows may be measured in individual units or family units. The subjective definition of y_i in such instances makes the definition of sample size subjective. Consequently, the use of significance tests, such as chi-square, in which the calculated value is sensitive to sample size but the critical value is not, leads to situations where the significance of a set of model estimates may be altered by simple redefinition of units.

(ii) We can often assume that a goodness-of-fit statistic has some theoretical distribution, but we do not know whether our assumption is valid in many instances. In some cases, for example, a statistic may be distributed according to a known theoretical distribution for large samples only, and we may have little idea as to its distribution when sample size is small, or indeed what constitutes a large or small sample.

5.2 Experimental Significance Tests

The use of theoretical distributions in significance testing is a legacy of late nineteenth and early twentieth century thinking when there was no other way to assess the significance of particular statistics. Today, however, we can take advantage of high-speed, large-memory computers and obtain experimental distributions for any statistic. In so doing, we do not need to make assumptions about the theoretical properties of particular statistics.

Two approaches to experimental significance testing exist: distributional methods (Openshaw, 1979; Diaconis and Efron, 1983) and randomisation methods (Kempthorne, 1955; Edgington, 1967; Hope, 1967; Costanzo, 1983). Distributional methods compare model estimates with those generated by random sampling from a known distribution. Randomisation methods seek comparison of model estimates with random arrangements of the observed data (Cliff and Ord, 1981). In both cases, a null hypothesis of no significant difference between the observed data $\underline{Y}$ and the set of model estimates $\hat{\underline{Y}}$ is rejected whenever the experimental procedure produces values superior to those obtained for the model in a specified number of cases (Hope, 1967; Openshaw, 1979). For example, to conclude that a model is adequate, one might require the experimental procedure (a random drawing of predicted values) to generate superior values of the statistic less than 5% of the time. The two experimental significance testing procedures are now discussed in greater detail.

5.2.1 Distributional Methods

Distributional methods are used to simulate the variance of a statistic when the distribution of the observed data (and hence the model estimates) is known, but the goodness-of-fit statistic has no known the theoretical distribution. When using distributional methods, the null hypothesis is that the model produces estimates that are not significantly different from a simple random drawing from the distribution that generated the observed data set. The significance testing procedure is as follows. A value of a goodness-of-fit statistic (chi-square, for example) is calculated using the observed data and the model estimates. Following this, samples (typically 99 or 999) of n cases each are drawn from an appropriate distribution, and a value of the goodness-of-fit statistic is calculated for each of the samples against the observed data set. These experimentally generated values of the goodness-of-fit statistic are then used to provide an experimental distribution for the value of the statistic determined with the model (Figure 7).

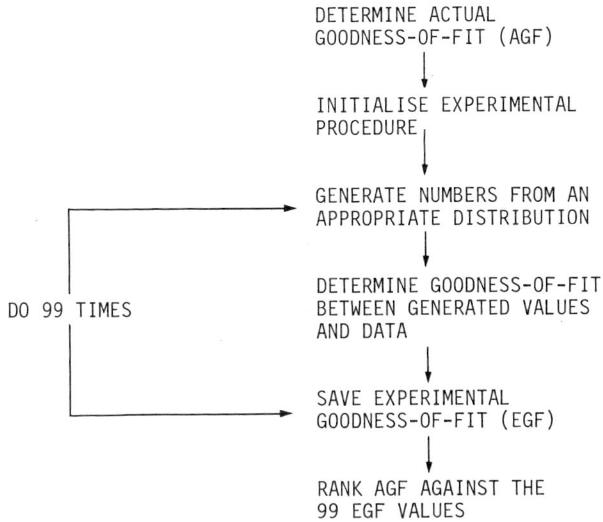

Figure 7. Distributional Experimental Variance Calculation

For example, for our sinkhole data the observed chi-square statistic is 9.3. These are count data drawn from a Poisson distribution with mean 5. To assess the significance of $\chi^2 = 9.3$, we draw 99 random samples from a Poisson distribution of mean 5, calculating a chi-square statistic for each of the 99 samples. If fewer than 5 of these random samples produce values of the chi-square statistic smaller than 9.3, then we can be at least 95% confident that our model is an accurate one. Using this experimental technique, we can generate Table 3, which gives the values of chi-square associated with each of the 99 samples. Since only two to these random samples are superior to the model estimates, we can conclude that the differences between the observed and predicted sinkhole occurrences are not significant, and that our model is an accurate one. The critical value of χ^2 in this experiment would be 12.47; values larger than this would yield five or more superior χ^2 values in the experimental distribution.

5.2.2 Randomisation Methods

Randomisation methods are used to simulate the variance of a statistic when the distribution from which we have drawn our sample is unknown. When using randomisation methods, the null hypothesis is that the value of the goodness-of-fit statistic associated with the model is not significantly different from that which would be expected from a random distribution of the observed values across the units of the study area. The significance

Table 3: Chi-Square Values generated by the Distributional Experimental Variance Method

12.47	11.25	23.44	17.19	23.23
47.75	22.79	41.09	38.73	121.19
17.32	87.73	38.36	25.73	36.20
116.45	63.05	23.65	17.17	34.14
83.29	46.68	19.16	18.50	123.98
19.15	17.90	17.21	22.95	64.60
36.08	52.79	45.15	62.72	31.95
41.51	15.38	39.57	20.66	7.47*
29.34	22.99	47.66	19.60	39.81
16.10	28.66	20.75	29.06	34.76
29.92	28.47	32.10	14.30	50.87
68.25	17.78	56.65	25.44	42.47
32.60	11.04	46.28	16.81	20.33
37.11	88.91	13.12	36.52	22.83
8.82*	15.60	22.68	15.36	20.38
30.75	101.31	108.23	37.84	38.03
36.27	26.07	39.99	12.89	22.42
55.01	17.51	16.23	30.14	43.89
17.14	19.78	39.11	32.49	42.41
51.42	41.12	27.01	30.19	

* values less than the actual goodness-of-fit

testing procedure is as follows. A value of a goodness-of-fit statistic, again let us use the example of chi-square, is calculated using the observed data and the model estimates. These observed data are then rearranged at random across the n units of the study area. These rearranged data are then compared with the model estimates and a chi-square statistic calculated. Alternatively, one could rearrange the model estimates and compare them with the observed values. This shuffling procedure is repeated for all n! possible assignments of data. For data sets with a large number of units, sampling with replacement from all n! possible combinations is used. This typically involves 99 or 999 samples. Following this shuffling procedure the values of the goodness-of-fit statistic are then used to provide an experimental distribution as before (Figure 8).

Again using our sinkhole example, chi-square is 9.3 and n is 10. Hence, there are 10! or 3 628 800 possible rearrangements of the observed data across the 10 sampling units. We will randomly sample 99 of these, calculating a chi-square statistic for each randomly drawn sample. If fewer than five of these randomly drawn samples produce values of chi-square less than 9.3, we can conclude that our model is accurate. Table 4 reports the chi-square values associated with 99 rearrangements of the sinkhole predictions. Coincidentally, we again draw only two superior values of chi-square, and hence conclude that our model is accurate. The critical value of X^2 here would be 12.92, which is very similar to the critical value reported for the distributional method.

An alternative method of deriving the critical value of chi-square from the randomisation procedure is to assume that the distribution of the chi-square values is asymptotically Normal (Mantel, 1967), and to compute

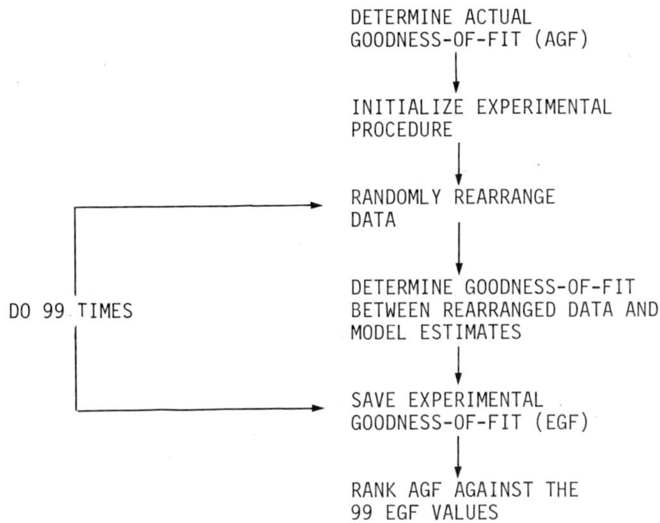

Figure 8. Randomisation Experimental Variance Calculation

Table 4: Chi-Square Values Generated by the Randomisation Experimental Variance Method

85.44	42.26	25.40	117.30	17.77
7.13*	48.70	60.18	66.30	18.97
39.90	115.23	55.49	33.43	43.38
68.75	34.47	76.88	84.68	13.50
55.92	82.95	12.17	37.44	58.05
84.35	76.17	21.67	83.17	52.40
25.24	24.35	104.42	4.65*	44.55
89.72	46.58	100.10	104.60	36.30
65.78	12.92	67.08	11.93	55.40
50.57	42.20	21.28	30.58	38.17
31.57	54.92	64.73	64.60	97.78
86.13	52.10	59.32	78.58	54.00
53.35	133.43	41.90	15.23	62.22
38.28	35.77	65.37	52.37	43.40
20.33	48.20	66.07	48.88	45.10
84.81	35.91	43.44	34.33	29.00
99.57	115.80	66.71	55.63	23.36
27.00	55.13	115.06	52.98	45.67
118.55	40.37	102.07	41.28	52.57
96.37	38.97	59.88	27.86	

* values less than the actual goodness-of-fit

critical values of the Normal distribution in the usual manner. That is, to use the standard deviation and mean of the experimental distribution to work backwards from the z-score at some critical point. To demonstrate this method, let us assume that the values of chi-square reported in Table 4 are drawn from a Normal distribution (we will examine this assumption in a moment). The mean of these values is 55.27, and the estimate of the population standard deviation is 28.94. If we postulate a null hypothesis, that the calculated value of chi-square (from our observed data) is not significantly less than the mean of that obtained from a random permutation of the data, then the region of rejection of this null hypothesis is all in the lower tail of the distribution. Hence, at the 95% significance level, the critical value of z is -1.645. Since z is defined by:

$$z = \frac{x_i - \bar{x}}{\sigma_x} , \qquad (37)$$

$$-1.645 = \frac{x_i - 55.27}{28.94} .$$

On rearranging, we obtain our critical value as:

$$x_i = (-1.645 \times 28.94) + 55.27 = 7.66 .$$

Hence, our calculated value of 9.33 lies within the region of acceptance, so that we cannot reject the null hypothesis. On this basis, we might not be satisfied with our model performance.

However, there are two problems with this latter procedure for assessing the critical value of a distribution derived from the randomisation procedure: one concerns the validity of the Normality assumption; the other concerns the use of the Normal distribution in this particular instance. Mielke (1979) points to the 'lumpiness' of permutational distributions, and argues that they converge to Normality only very slowly as the number of permutations increases, and that in any particular instance the fit to Normality may not be very accurate. To investigate this criticism, we can examine the fit of the distribution of values given in Table 4 to Normality with a Kolmogorov - Smirnov test (Taylor, 1977).

By categorising the distribution by z-scores, we can present a visual representation of the fit of the observed distribution to that of a perfect Normal one (see Figure 9). The fit appears to be a reasonably close one, although there are some noticeable deviations. In particular, the observed distribution seems to contain more values that are below the mean than would be expected in a Normal distribution. The observed and expected relative frequencies for each of these divisions and the cumulative observed and expected frequencies are given in Table 5. One goodness-of-fit test for Normality that is appropriate here utilises the Kolmogorov-Smirnov statistic, D, which is defined as:

$$D = \max|C_O - C_E| , \qquad (38)$$

where C_O is the cumulative observed frequency, and C_E is the cumulative expected frequency. For our data, D = 0.0757 (for the interval 40.8 to 55.27), which lies between the critical values of D when testing goodness-of-fit to a Normal distribution of 0.074 (α=.2) and .077 (α=.15). Again, this is a test where we should be concerned with Type II error and hence we should use large values of α. We do not want to make it easy to accept the null

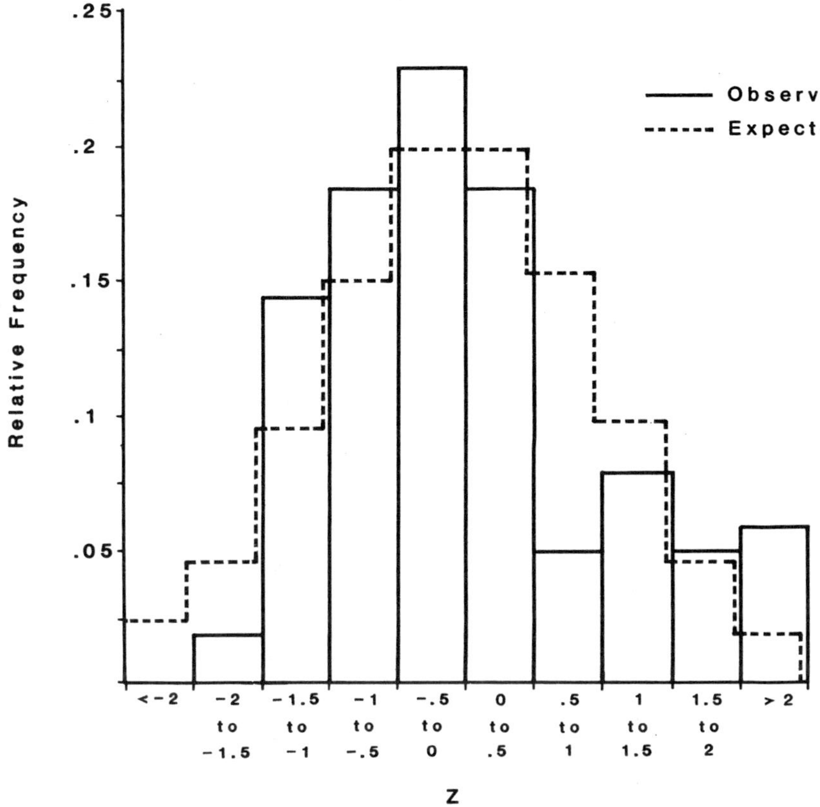

Figure 9. Observed and expected (under assumption of normality) Frequency Distribution of Chi-square values obtained from the randomisation method

hypothesis that there is no significant difference between our observed distribution and a Normal distribution. Hence, depending upon the stringency of our test, we may or may not reject this null hypothesis. To be consistent with the rest of this text, we should select α =0.25 here and hence we would reject the null hypothesis and conclude that the distribution of chi-square values reported in Table 4 is significantly non-Normal. Consequently, our assumption of Normality used to derive the critical chi-square value is invalid.

The second criticism of the use of the Normal distribution in this instance applies even if the permutational values had a Normal distribution. The Normal distribution is unbounded in both tails, whereas the lower bound of the chi-square statistic is zero. Consequently, the use of the Normal distribution in this instance makes it very difficult to reject the null

Table 5: Frequency Distribution of Chi-Square Values Obtained from the Randomisation Method

Division in z scores	Division in Terms of Table 4	Observed Number of Occurrences	Observed Relative Frequency	Expected Relative Frequency	Cumulative Observed Frequency	Cumulative Expected Frequency
< -2	< -2.61	0	0	.0228	0	.0228
-2 to -1.5	-2.61 to 11.86	2	.0202	.0440	.0202	.0668
-1.5 to -1	11.86 to 26.33	14	.1414	.0919	.1616	.1587
-1 to -0.5	26.33 to 40.8	18	.1818	.1498	.3434	.3085
-0.5 to 0	40.8 to 55.27	23	.2323	.1915	.5757	.5
0 to 0.5	55.27 to 69.74	18	.1818	.1915	.7575	.6915
0.5 to 1	69.74 to 84.21	5	.0505	.1498	.8080	.8413
1.0 to 1.5	84.21 to 98.68	8	.0808	.0919	.8888	.9332
1.5 to 2	98.68 to 113.15	5	.0505	.0440	.9393	.9772
> 2	> 113.15	6	.0606	.0228	1.0	1.0

hypothesis and conclude that our model is an accurate one. In some cases, it may even be impossible to reject the null hypothesis. Consider in our example if we had chosen a significance level of α =0.025 instead of α =0.05. The critical chi-square value under the assumption of Normality would then be:

$$(-1.96 \times 28.94) + 55.27 = -1.45 \ ,$$

which is an impossible chi-square value. Consequently, the assumption of Normality in this instance is not recommended.

6. SPATIAL ASPECTS OF GOODNESS-OF-FIT

While much of the previous discussion has been relevant to goodness-of-fit testing in both spatial and non-spatial situations, two points that are peculiar to spatial goodness-of-fit are now described briefly. The first is that of difference maps and tests of spatial autocorrelation on residuals; the second is that of limitations on the range of goodness-of-fit statistics due to spatial restrictions.

6.1 Difference Maps and the Spatial Autocorrelation of Regression Residuals

At the beginning of this monograph, in Figure 3, we presented a map of the residuals from our income model. From a visual inspection of this map we assumed that there was no significant spatial autocorrelation, so that our regression was valid and we had not omitted any relevant variable from the model. We can examine this assumption more objectively by calculating a spatial autocorrelation statistic for the residuals (see Cliff and Ord, 1981 for a discussion of spatial autocorrelation).

The test for spatial autocorrelation among regression residuals is rather complex and demands an understanding of spatial autocorrelation that cannot be covered here. Hence, we can do little more than refer to the appropriate parts of Cliff and Ord and leave the reader to pursue this topic. The test statistic for spatial autocorrelation among regression residuals is the Moran coefficient, defined as:

$$I = \frac{n \sum\sum_{ij} w_{ij} \, e_i \, e_i}{W \sum\sum_{ij} e_i^2} \ , \qquad (39)$$

where n is the number of residuals, w_{ij} is the weight (strength of the link) between region i and region j, e_i is $y_i - \hat{y}_i$, e_j is $y_j - \hat{y}_j$, and $W = \sum\sum_{ij} w_{ij} \, (j \neq i)$ The definition of the weighting scheme is somewhat subjective, although two commonly employed weighting schemes are:

(i) w_{ij} = 1 if regions i and j share a common edge
= 0 otherwise

(ii) $w_{ij} = 1/d_{ij}$ where d_{ij} is the distance between the centroids of regions i and j.

For the residual map in Figure 3, using the weighting scheme in (i), $I = -0.0830$. The value of I has extremes of +1 when there is perfect positive spatial autocorrelation (positive residuals are associated with positive errors in neighbouring regions and negative residuals are associated with negative errors in neighbouring regions), and -1 when there is perfect negative spatial autocorrelation (positive residuals are associated with negative residuals in surrounding regions and vice-versa). When there is no spatial autocorrelation, the statistic has a value of approximately zero. Here then, the value of $I = -0.0830$ appears to indicate the presence of only weak spatial autocorrelation. However, we examine the significance of this value by calculating a standard Normal deviate:

$$z = \frac{I - E(I)}{Var(I)} \quad . \tag{40}$$

The expressions for E(I) and Var(I) are given in Cliff and Ord (1981) as equations (8.21) and (8.29), respectively. The particular form of these equations depends on the number of parameters estimated in the regression equation from which the model predictions are derived. For our model of predicted mean incomes where $n = 20$ and $I = 0.0587$, $E(I) = -0.0717$ and $\sqrt{Var(I)} = 0.1200$, so that $z = 1.0874$. The critical value of z at the 95% confidence level (the standard confidence level is appropriate here since we want to minimise Type I error) is 1.96, so that we can accept the null hypothesis that there is no significant spatial autocorrelation in the residuals. Hence, we have no reason to suspect that our model calibration technique (ordinary least squares regression) was inappropriate, or that our model is poorly specified.

6.2 Spatial Restrictions on the Range of Goodness-of-Fit Statistics

When investigating the goodness-of-fit between certain types of spatial data, it is sometimes impossible, due to spatial constraints, to obtain values of the goodness-of-fit statistic within the complete theoretical range of the statistic. It is also possible that the expected value of the statistic in practice is different from the theoretical expected value. Consider, for example, an investigation of the relationship between access to health-care provision and income, where mean incomes are obtained for census tracts within a city, and access to health-care provision is measured by the distance from the centroid of each census tract to the nearest health-care facility. Suppose we take the example of the income data mapped in Figure 1. These data are relatively fixed, and it is assumed in the model structure that the location of health-care facilities is determined in part by this spatial pattern of incomes. However, given the income distribution in Figure 1, there is no spatial arrangement of health-care facilities that will yield a value of $R^2 = 1$ in our regression model. Any spatial arrangement of facilities will yield a set of distances that are not perfectly correlated with income. McLafferty (1982) demonstrates, for example, that in Cedar Rapids, Iowa, the range of R for the above two variables is approximately -0.4 to 0.8 with a mean of approximately 0.4. These values are obtained by taking a large number of random permutations of the locations of the health-care facilities and calculating R for each permutation. A similar problem has been noted by Joseph (1982) with the range of the coefficient of localisatior a representative of the family of general distance statistics discussed in Section 3.2.

In this kind of situation, which must occur frequently in geographical studies, it would be more useful to assess the significance of a particular goodness-of-fit statistic with reference to an experimental distribution, as described in Section 5, rather than a theoretical distribution.

7. GOODNESS-OF-FIT TESTS IN DISCRETE CHOICE MODELLING

The modelling of discrete choices is a branch of a broader research area known as categorical data analysis, the use of which is expanding within geography (Wrigley, 1982, 1985). It is increasingly being realized that while many of the choices individuals make are continuous (such as how much to pay for a home, or how much fertilizer to apply to a crop), and can therefore be modelled by standard techniques such as ordinary least squares regression, many choices are discrete and need specialised modelling techniques. Examples of such techniques include the logit model (McFadden, 1974), the probit model (Daganzo, 1979), the dogit model (Gaudry and Dagenais, 1979), and the elimination-by-aspects model (Tversky, 1972).

Three goodness-of-fit statistics are commonly reported in applications of discrete choice models. One of these is the deviance which we have already discussed. The other two are the rho-square statistic and Wolfe's correlation coefficient which we will now discuss briefly. A fourth goodness-of-fit statistic that we will examine is Kendall's tau, which is not used as frequently as the above statistics in discrete choice modelling, but which is used in questionnaire analysis.

The rho-square statistic is actually a likelihood statistic which is closely related to the deviance measure. It is defined as:

$$\rho^2 \doteq 1 - (\hat{L}^*/\hat{L}(0)^*) \quad , \tag{41}$$

where $\hat{L}^*$ represents the log-likelihood associated with the calibrated model (see Section 3.4), and $\hat{L}(0)^*$ represents the log-likelihood associated with a model where all the parameters except the constant are zero. The statistic has a minimum at zero when the model is no more accurate than the mean and hence when $\hat{L}^* = \hat{L}(0)^*$. It has a maximum that tends to one as the model becomes increasingly more accurate than the mean. To see that the maximum tends to one, it is necessary to realize that both $\hat{L}^*$ and $\hat{L}(0)^*$, being log-likelihoods of probabilities, are less than zero and hence, as the model becomes increasingly more accurate than the mean, $\hat{L}^*$ tends to zero from the negative side while $\hat{L}(0)^*$ is a large negative number. The ratio $\hat{L}^*/\hat{L}(0)^*$ is always less than or equal to one and positive, and therefore tends to zero from the positive side so that ρ^2 tends to one.

While this statistic is commonly reported in discrete choice modelling applications, two problems exist with its use. One is that no theoretical significance testing procedure is available; the other is that ρ^2 is relatively insensitive to variations in model performance. In practice it is rare to see values outside the range $0.1 < \rho^2 < 0.4$.

One other goodness-of-fit statistic that is becoming increasingly popular in applications of discrete choice models, although it could be applied to any model, is a correlation coefficient proposed by Wolfe (Wolfe, 1976) and extended by Hubert and Golledge (1981). The statistic is used to assess the relative merits of two models. If A represents a set of observed choices or proportions, and B and C represent predictions of these choices from two separate models, then the correlation coefficient:

$$r_{A(B-C)} = \frac{n \sum_i a_i(b_i - c_i) - \sum_i a_i \cdot \sum_i (b_i - c_i)}{\left\{[n \sum_i a_i^2 - (\sum_i a_i)^2]\,[n \sum_i (b_i - c_i)^2 - \sum_i (b_i - c_i)]^2\right\}^{\frac{1}{2}}}$$

$$= \frac{r_{AB} - r_{AC}}{[2(1 - r_{BC})]^{\frac{1}{2}}} , \qquad (42)$$

describes the difference between B and C, and hence between the two models generating the predicted matrices. As with the deviance measure previously discussed, one of these sets of predictions could come from a 'full' model; that is, one that exactly replicates the observed data. The values a_i, b_i and c_i are elements of the sets A, B, and C, respectively, and the statistics, r_{AB}, r_{AC} and r_{BC} are the simple correlation coefficients between A and B, A and C, and B and C, respectively. If $r_{A(B-C)}$ is significantly different from zero, in either direction, then the null hypothesis that the models yielding the predictions in B and C are equivalent can be rejected.

The right-hand expression of equation (42) clearly demonstrates that $r_{A(B-C)}$ is a measure of the difference between two dependent correlations, r_{AB} and r_{AC}. If the null hypothesis is rejected in the positive tail of the distribution, this suggests that B is a significantly more accurate set of predictions than C; if the null hypothesis is rejected in the negative tail, this suggests that C is a significantly more accurate set of predictions than B. The assessment of the significance of $r_{A(B-C)}$ can be undertaken with the t-test described in equation (7), or by deriving an experimental distribution of the statistic by randomly permuting the predicted (B-C) data set as described in Section 5. However, as Longley (1984) points out, there is a weakness in relying on an experimental distribution in this instance, because there is no way of accounting for different degrees of freedom in the two models being compared. That is, the significance of a difference in model performance cannot be assessed accurately if the two models being compared have different numbers of parameters. Models with a greater number of parameters are expected to perform more accurately, *ceteris paribus*: we often need to assess whether the improvement in model performance due to the addition of parameters is significant, and this cannot be done with experimental significance testing procedures. Of course, if a model with more parameters does not perform more accurately than one with fewer, we should, on the grounds of parsimony, select the latter.

Two prerequisites for the use of this correlation coefficient are that $r_{BC} = 1$ and that the variances of B and C are equal. The latter is usually assured by standardizing all three data sets in some way, such as by taking z-scores or by deriving proximity matrices. Longley (1984) and Halperin *et al.*, (1984) provide examples of the latter technique in geographic modelling.

To see the application of the goodness-of-fit statistic in equation (42), consider the following three data sets taken from Longley (1984); in this case we will take $\underline{A}$ to represent the observed choices of six individuals between two residential locations, one in the center of the city, and the other in the suburbs. $\underline{B}$ and $\underline{C}$ are the predictions of these choices derived from two different discrete choice models. The task is to determine whether or not there is a significant difference between the two sets of predictions.

		$\underline{A}$ choices 1	$\underline{A}$ choices 2	$\underline{B}$		$\underline{C}$	
individuals	1	1	0	0.4	0.6	0.5	0.5
	2	1	0	0.7	0.3	0.6	0.4
	3	1	0	0.5	0.5	0.7	0.3
	4	0	1	0.2	0.8	0.1	0.9
	5	1	0	0.9	0.1	0.8	0.2
	6	0	1	0.6	0.4	0.4	0.6

In this particular instance, the $\underline{B}$ and $\underline{C}$ matrices do not need to be normalised because they have the same variance (they are merely the same numbers in a different order). Hence, the $(\underline{B}-\underline{C})$ matrix can be computed directly as:

$(\underline{B}-\underline{C})$	
-0.1	0.1
0.1	-0.1
-0.2	0.2
0.1	-0.1
0.1	-0.1
0.2	-0.2

,

and $r_{A(B-C)} = -0.47$. The negative value of the correlation coefficient suggests that matrix $\underline{C}$ contains a more accurate set of predictions than matrix $\underline{B}$. However, from equation (7), t=1.065 and since the critical value of t is 2.78 we cannot reject the null hypothesis that there is no significant difference between the $\underline{B}$ and $\underline{C}$ matrices. A more detailed application of this statistic and its experimental distribution are given by Longley (1984) in a comparison of several models of residential choice.

A goodness-of-fit statistic that has been employed less frequently in discrete choice modelling, but that can be used whenever the observed data and the model predictions consist of a set of ranked values, is Kendall's tau. In discrete choice modelling such data are often preference rankings for various alternatives. However, perhaps a more common geographic situation, where the data are ranked or ordinal, occurs when individuals are asked to respond to statements from a questionnaire on a scale such as: strongly agree, agree, neutral, disagree, strongly disagree.

To see the application of Kendall's tau, consider the following experiment which is very similar to that described by Timmermans (1984). An individual is asked to give his/her preferences for 9 hypothetical shopping destinations, each of which has a certain combination of two attributes: the choice of goods (as determined by store size), and travel time from the individual's residence. Both attributes have three levels: choice of goods can be limited, medium or wide; travel time can be 15, 30, or 45 minutes. Each of the 9 hypothetical shopping destinations has a different combination of the two attributes. The preferences of the individual are ranked from 1 (most preferred destination) to 9 (least preferred). A model is then postulated with the following form:

$$y_i = \sum_{k=1}^{K} \sum_{m_k=1}^{M_k} \beta_{k m_k} x_{ikm_k} + e_i \quad , \qquad (43)$$

where y_i is the ranked preference of the individual for the ith shopping destination; k represents a particular attribute of which there are K; m_k represents the level of the kth attribute; x_{ikm} = 1 if destination i has attribute k at the m_kth level and = 0 otherwise; β_{km} is a parameter to be estimated that relates y_i and x_{ikm}; and e_i is an error term.

The model, when calibrated against data on y_i , is used to obtain estimates of y_i which are converted to a ranking scale similar to y_i, and a goodness-of-fit statistic is needed to compare $\hat{\underline{Y}}$ and $\underline{Y}$. In this case, both $\hat{\underline{Y}}$ and $\underline{Y}$ will consist of a set of ranked data, and Kendall's tau is the appropriate goodness-of-fit statistic. The statistic is the difference in the proportions of concordant and discordant pairs of rankings, and can be written as:

$$\tau = \frac{C}{n(n-1)/2} - \frac{D}{n(n-1)/2} \quad , \qquad (44)$$

where C represents the number of concordant pairs of observations, D represents the number of discordant pairs of observations, and n represents the number of rankings, so that n(n-1)/2 is the total number of pairs of observations. A pair of observations is said to be concordant if the observation that ranks higher of the two in $\underline{Y}$ also ranks higher of the two in $\hat{\underline{Y}}$ (that is, the order of the two rankings is the same in both sets). A pair of observations is said to be discordant if the observation that ranks higher of the two in $\underline{Y}$ ranks lower of the two in $\hat{\underline{Y}}$ (that is, the order of the rankings is reversed).

When there is no association between $\hat{\underline{Y}}$ and $\underline{Y}$, C = D and $\tau = 0$. When $C > D$ and $\tau > 0$, there is a positive relationship between the rankings in $\hat{\underline{Y}}$ and $\underline{Y}$; when $C < D$ and $\tau < 0$, there is an inverse relationship between the rankings in $\hat{\underline{Y}}$ and $\underline{Y}$. When all pairs are concordant (a perfect model fit), $\tau = +1$; when all pairs are discordant (high ranking predictions are associated with low actual rankings and vice versa), $\tau = -1$.

As an example of the application of Kendall's tau, consider the data on 9 shopping destination preferences in Table 6.

The concordant pairs are:

(D1,D3), (D1,D4), (D1,D5), (D1,D6), (D2,D3), (D2,D4), (D2,D5), (D2,D6), (D2,D7), (D3,D4), (D3,D5), (D3,D6), (D3,D7), (D3,D8), (D3,D9), (D4,D5), (D4,D6), (D4,D7), (D4,D8), (D4,D9), (D5,D7), (D5,D8), (D5,D9), (D6,D7), (D6,D9) .

Table 6: Data on Shopping Preferences to Derive Kendall's Tau

Shopping Destination	Stated Preference Ranking	Predicted Preference Ranking
D1	5	7
D2	6	5
D3	1	1
D4	9	9
D5	2	3
D6	4	2
D7	7	6
D8	3	8
D9	8	4

The discordant pairs are:

(D1,D2), (D1,D7), (D1,D8), (D1,D9), (D2,D8), (D2,D9), (D5,D6), (D6,D8), (D7,D8), (D7,D9), (D8,D9) .

Hence, C = 25, D = 11, and n = 9 so that:

$$\tau = \frac{25}{36} - \frac{11}{36} = 0.389 \quad ,$$

which indicates a positive association between $\hat{\underline{Y}}$ and $\underline{Y}$. To examine the significance of this value, the statistic:

$$z = \tau / \sqrt{\frac{2(2n+5)}{9n(n-1)}} \tag{45}$$

is approximately Normally-distributed. In this example:

$$z = 0.389 \; / \sqrt{46/648} \; = \; 1.46 \quad .$$

Here, the minimum value of z indicates a perfect model fit, so that to be conservative in accepting the null hypothesis that the differences between $\hat{\underline{Y}}$ and $\underline{Y}$ are not significant and that we have an acceptable model, we should use a confidence level much lower than the traditional 95% or 99% levels. We select a 75% confidence level at which the critical value of z is 1.15. Consequently, at this level we reject our null hypothesis that there is no significant difference between $\hat{\underline{Y}}$ and $\underline{Y}$ and conclude that our model is not particularly accurate.

For further details on the use of Kendall's tau, especially in the case where tied rankings are present, see Agresti and Agresti (1979). Timmermans (1984) provides a further empirical application in geography.

8. APPLICATIONS

Up to this point, we have examined a variety of goodness-of-fit statistics, and have discussed the theoretical and computer-generated approaches to assessing the significance of each of these measures. Although we have provided worked examples throughout, we now provide a review of several interesting applications of goodness-of-statistics in geographic literature. We draw our examples mainly from spatial interaction modelling, input-output analysis, and climatological modelling.

Examples of the use of goodness-of-fit statistics in spatial interaction modelling include Fotheringham's (1983) use of R^2 to judge differences in linear interaction model specification; he concluded that competing destinations models are superior to traditional gravity models. Lewis (1975) similarly uses R^2 to judge differences in model performance between gravity and Heckscher-Ohlin models of interregional trade. Baxter and Ewing (1979, 1981) use χ^2 to discriminate between several alternative models of recreational trip behavior.

Use of generalised distance measures in the spatial interaction modelling literature includes the use of RMSE by Black (1973) to assess goodness-of-fit in models of interregional commodity flow. Hathaway (1975) uses two general distance statistics as well as χ^2 to assess the fit of interaction models disaggregated by age, sex, martial status, socioeconomic group, occupational class, and standard industrial class relative to an aggregate model. Pitfield (1978) uses SRMSE to discriminate between a linear programming model and a doubly-constrained gravity model of freight movements in Britain.

Examples of the use of information-based and likelihood statistics within the spatial interaction modelling literature include Ayeni's (1982) use of psi to assess goodness-of-fit in models of commuting in Lagos, Nigeria, and Thomas' (1977) study of journeys to work in Merseyside. The latter study is particularly interesting because Thomas uses the additive properties of information gain to examine the interrelationship between model fit and spatial structure. Southworth (1983) uses a likelihood measure to assess goodness-of-fit in models which describe the interrelationships between spatial structure and distance decay.

Other examples of the use of goodness-of-fit statistics in spatial interaction modelling include Fotheringham and William's (1983) use of three statistics (R^2, a general distance statistic, and information gain) to compare Poisson gravity models with log-normal and production-constrained models in four different data sets. The three statistics have the same order of magnitude in all four data sets, indicating a fair degree of correspondence between the statistics.

Lovett, Whyte and Whyte (1985) also examine Poisson gravity models, in their case with data on historical migration in Scotland. They use the percentage change in the deviance measure, which is reported as part of the GLIM package, to identify both relevant explanatory variables of the migration process and the most appropriate model structure.

The previously cited articles use goodness-of-fit statistics either as simple relative measures of model performance or with reference to traditional significance tests. The seminal paper illustrating the use of computer-generated significance testing is Gale *et al.* (1984), which uses randomisation methods to discriminate between nine different specifications of an interregional migration model.

A second body of the geographic literature in which goodness-of-fit is an area of debate is input-output analysis. Butterfield and Mules (1980) provide an excellent review of goodness-of-fit issues in input-output modelling as well as an example of nonparametric methods, X^2, R^2, and t, and MAD (mean absolute difference, a general distance statistic) to assess estimates of the Western Australian input-output table. Kim, Boyce and Hewings (1983) use X^2, R^2 and t to assess the fit of a combined input-output and commodity flow model for the Korean economy. McMenamin (1973) uses X^2 and MAD to assess different survey methods for estimating the Washington State input-output table. Similarly, Schaffer and Chu (1969) and Czamanski and Maliza (1969) investigate several survey methods for the Washington State input-output table. Shaffer and Chu utilize X^2 to chose between methods, while Czamanski and Maliza use MAD and a form of information gain.

A third body of literature where the use of goodness-of-fit statistics is prevalent involves climatological modelling. Here Willmott (1984) serves as an excellent review and also provides examples of generalised distance measures, particularly RMSE, to assess alternative evapotranspiration models. Suckling and Hay (1976) use RMSE to discriminate between alternative models of solar radiation tested on three locations in Canada, while Atwater and Bell (1978) utilize the correlation coefficient, mean absolute deviation and RMSE to assess the reliability of a numeral solar radiation model. Willmott and Weeks (1980) use R^2 and RMSE to assess the validity of a rainfall interpolation model applied to California, while Johnson and Bras (1978) use RMSE to assess the fit of a model of short-term rainfall rates using simulated data as well as actual data on these rates from stations in Oklahoma and Illinois. Burt *et al.* (1980) use RMSE to assess the predictive abilities of a simulated water balance model for irrigated and rain-fed agricultural crop production using data from four sites in the southwestern United States. Finally, Powell *et al.* (1984) use three statistics, mean absolute deviation, RMSE, and R^2 to examine models used to estimate solar radiation from satellite imagery. Interestingly, they use RMSE to compare model performance in several data sets, although this unstandardised version is data specific. The SRMSE should be used for comparative purposes.

Lastly, we note an important study by Openshaw and Taylor (1982), who examine the effects of zonal definition on the correlation between the spatial distribution of Republican voters and that of the elderly in Iowa. While not explicitly concerned with goodness-of-fit, since the topic of the paper is on the correlation between two variables rather than on the modelling of one particular variable, the paper nevertheless has important implications for goodness-of-fit statistics. Openshaw and Taylor demonstrate the sensitivity of spatial statistics (and this would include spatial goodness-of-fit statistics) to zonal definition to the extent that different aggregations of 99 zones to 6 zones yielded correlation coefficients as diverse as -0.99 and +0.99! This suggests that we should be very careful about generalising the results of goodness-of-fit tests whenever models are calibrated with aggregate zonal data.

9. CONCLUSIONS

We have analysed the performance of several goodness-of-fit statistics with respect to their overall performance, their sensitivity to error and their use in significance testing. The results of this analysis are summarised in Table 7. For significance testing, theoretical distributions are

Table 7: Summary of Goodness-of-fit statistics evaluated

Statistic	Minimum Value	Maximum Value	Minimum Value Indicates	Sensitivity to Error
R^2	0	1	No Correspondence	Non-Linear
Pearson X^2	0	infinity	Perfect Correspondence	Non-Linear
SRMSE	0	1 (usually)	Perfect Correspondence	Linear
Information Gain	0	infinity	Perfect Correspondence	Non-Linear
Psi	0	n 1n 2	Perfect Correspondence	Non-Linear
AED	0	1n n	Perfect Correspondence or same values in different order	Non-Linear
Deviance	0	infinity	Perfect Correspondence	Non-Linear

Table 7 - continued

Theoretical Distribution Available	Experimental Distribution Available	Comments
Yes, but very insensitive	Yes	Measures linear correlation. Only appropriate for models with Normally-distributed errors
Yes, but problems with definition of units	Yes	Sensitive to underpredictions. Not standardised for units. Only appropriate for models with Poisson errors.
No	Yes	Statistic can be greater than one if average error exceeds mean value
Yes, MDI statistic is assumed to be chi-square distributed	Yes	Emphasises errors in large values. Problems with definitions of units. Cannot have predicted values equal to zero.
Yes, MDI statistic is assumed to be chi-square distributed	Yes	More conservative test than information gain. Can have predicted values equal to zero.
Yes, t distribution or Normal distribution. However, doubts exist regarding validity of variance formulation	Yes	Can falsely conclude model is accurate when values are the same in two data sets but appear in different order.
Yes, chi-square (for Normal errors) or assumed chi-square distributed (for Poisson errors)	Yes	For Poisson errors, it is identical to MDI. Useful for comparing the performance of two models.

available for most of the statistics examined here, although experimental methods can be used for all statistics. These latter methods are particularly useful when asymptotic relationships fail to hold, when ambiguity exists about sample size, or when statistical assumptions cannot be met. Experimental methods additionally provide extremely intuitive ways of assessing model goodness-of-fit.

While we offer no single solution to the problem of goodness-of-fit assessment, we caution against the dogmatic acceptance of the results of goodness-of-fit tests. For example, Ayeni (1982), while concluding that his model is accurate, reports values for the MDI statistic that could represent errors in excess of 100%. Similarly, Thomas (1977) reports values of information gain, ranging from 0.30 to 0.35, that could be indicative of error levels in excess of 80%. In these instances, problems arise because of the sensitivity of X^2 distributions to sample size, and because there is no good understanding of the practical range of statistics such as information gain.

It is doubtful that a single ideal goodness-of-fit statistic exists. However, we have provided a systematic framework for experimenting with statistics to determine their usefulness in particular instances. Given a correct definition of error and the definition of an ideal statistic (in terms of the properties a statistic must possess in a particular instance), it is possible to assess the response surfaces of statistics and their associated tests with respect to the ideal. An awareness of how statistics behave in particular instances should lead to a more careful and practical use of goodness-of-fit statistics in geographic research.

REFERENCES

A. THEORY

Abraham, A.D. and D.M. Mark (1986), The random topology model of channel networks: bias in statistical tests. *The Professional Geographer*, 38, 77-81.

Agresti, A. and B. Finlay-Agresti (1979), *Statistical methods for the social sciences*, (San Francisco: Dellen).

Anselin, L. (1984), Specification tests and model selection for aggregate spatial interaction: an empirical comparison. *Journal of Regional Science*, 24, 1-16,

Bishop, Y.M.M., S.E. Feinberg and P.W. Holland (1975), *Discrete multivariate analysis: theory and practice*, (Cambridge: MIT Press).

Black, J.A. and R.T. Salter (1975), A statistical evaluation of the accuracy of a family of gravity models. *Proceedings, Institution of Civil Engineers*, Part 2, 59, 1-20.

Butterfield, M. and T. Mules (1980), A testing routine for evaluating cell-by-cell accuracy in short-cut regional input-output tables. *Journal of Regional Science*, 20, 293-310.

Cliff, A.D. and J.K. Ord (1981), *Spatial processes: models and applications*, (London: Pion).

Constanzo, C.M. (1983), Statistical inference in geography; modern approaches spell better times ahead. *Professional Geographer*, 35, 158-165.

Constanzo, C.M. and N. Gale (1984), Evaluating the similarity of geographic flows. *Professional Geographer*, 36, 182-187.

Daganzo, C.F. (1979), *Multinomial Probit: the theory and its application to demand forecasting*, (New York: Academic Press).

Diaconis, P. and B. Efron (1983), Computer-intensive methods in statistics. *Scientific American*, 5, 116-130.

Duncan, O.D. and B. Duncan (1955), A methodological analysis of segregation indices. *American Sociological Review*, 20, 210-217.

Edgington, E.S. (1969), *Statistical inference: the distribution-free approach*, (New York: McGraw-Hill).

Ferguson, R. (1977), *Linear Regression in Geography*. CATMOG 15; (Norwich: Geo Abstracts).

Gaudry, M.J.I. and M.G. Dagenais (1979), The Dogit Model. *Transportation Research*, B 13B, 105-111.

Hanushek, E.A. and T.E. Jackson (1977), *Statistical methods for social scientists*, (New York: Academic Press).

Hope, A.C.A. (1968), A simplified Monte Carlo significance test procedure. *Journal of the Royal Statistical Society, Series B*, 30, 582-598.

Hubert, L.J. and R.G. Golledge (1981), A heuristic method for the comparison of related structures. *Journal of Mathematical Psychology*, 23, 214-226.

Hutcheson, K. (1970), A test for comparing diversities based on the Shannon formula. *Journal of Theoretical Biology*, 29, 151-154.

Johnston, R.J. and R.K. Semple (1983), *Classification using information statistics*, CATMOG 37; (Norwich: Geo Abstracts).

Joseph, A.E. (1982), On the interpretation of the coefficient of localisation. *Professional Geographer*, 34, 443-446.

Kempthorne, O. (1955), The randomization theory of experimental inference. *American Statistical Association Journal*, 61, 946-967.

Knudsen, D.C. and A.S. Fotheringham (1986), Matrix comparison, goodness-of-fit and spatial interaction modelling. *International Regional Science Review*, 10, 127-147.

Kullback, S. (1959), *Information theory and statistics*, (New York: John Wiley and Sons).

Kullback, S. and R.A. Leibler (1951), On information and sufficiency. *Annals of Mathematical Statistics*, 22, 78-86.

McCullagh, P. and J.A. Nelder (1983), *Generalized linear models*, (London: Chapman and Hall).

McFadden, D. (1974), Conditional logit analysis of qualitative choice behavior. In: *Frontiers in Econometrices*, (ed) P. Zarembka, (New York: Academic Press).

McLafferty, S. (1982), Urban structure and geographical access to public services. *Annals of the Association of American Geographers*, 72, 347-354.

McLafferty, S.L. and A. Gosh (1982), Issues in measuring differential access to public services. *Urban Studies*, 19, 383-390.

Mantel, N. (1967), The detection of disease clustering and a generalized regression approach. *Cancer Research*, 27, 209-220.

Mielke, P.W. (1979), On asympotic non-normality of null distributions of MRPP statistics. *Communications in Statistical Theory and Methods*, A8 (15), 1541-1550.

Nelder, J.A. and R.W.M. Wedderburn (1972), Generalised linear models. *Journal of the Royal Statistical Society*, 135, 370-384.

Openshaw, S. and P.J. Taylor (1982), A million or so correlation coefficients: three experiments of the mofifiable areal unit problem. In: *Statistical Applications in the Spatial Sciences*, (ed) N. Wrigley, (Norwich: Pion).

Phillips, F.Y. (1981), *A guide to MDI statistics for planning and management model building*, (Austin: Institute for Constructive Capitalism).

Pickles, A. (1986), *An introduction to likelihood analysis*, CATMOG 42; (Norwich: Geo Abstracts).

Pielou, E.C. (1969), *An introduction to mathematical ecology*, (New York: John Wiley).

Pindyck, R.S. and D.L. Rubinfield (1976), *Econometric models and economic forecasts*, (New York: McGraw-Hill).

Shannon, C.E. (1948), A mathematical theory of communication. *The Bell System Technical Journal*, 27, 379-623.

Silk, J. (1979), *Statistical concepts in geography*, (London: Allen and Unwin).

Smith, D.P. and B.G. Hutchinson (1981), Goodness-of-fit statistics for trip distribution models. *Transportation Research*, 15A, 295-303.

Snickars, F. and T.W. Weibull (1977), A minimum information priciple: theory and practice. *Regional Science and Urban Economics*, 7, 137-168.

Taylor, P.J. (1977), *Quantitative methods in geography*, (Boston: Houghton Mifflin).

Thomas, R.W. (1981), *Information statistics in geography*, CATMOG 31; (Norwich: Geo Abstracts).

Timms, D.W.G. (1966), Quantitative techniques in urban social geography. In: *Frontiers in Geographical Techniques*, (eds) R.J. Chorley and P. Haggett, (London: Methuen).

Tribus, M. and Rossi (1973), On the Kullback information measure as a basic for information theory: comments on a proposal by Hobson and Cheng. *Journal of Statistical Physics*, 9, 395-417.

Tversky, A. (1972), Elimination-by-aspects: a theory of choice. *Psychological Review*, 79, 281-299.

Upton, G.J.G. and B. Fingleton (1979), Log-linear models in geography. *Transactions, IBG*, New Series 4, 103-115.

Vincent, P., and J. Haworth (1984), Statistical inference: the use of the likelihood function. *Area*, 16, 131-146.

Willemain, T.R. (1980), *Statistical methods for Planners*, (Cambridge, Mass.: The MIT Press).

Willmott, C.T. (1984), On the evaluation of model performance in physical geography. In: *Spatial Statistics and Models*, (eds) G.L. Gaile and C.T. Willmott, (New York: Reider).

Wilson, S.R. (1976), Statistical notes on the evaluation of calibrated gravity models. *Transportation Research*, 10, 343-345.

Wolfe, D.A. (1976), On testing equality of related correlation coefficients. *Biometrika*, 63, 214-215.

Wrigley, N. (1982), Quantitative methods: developments in discrete choice modelling. *Progress in Human Geography*, 6, 547-562.

Wrigley, N. (1985), *Categorical data analysis for geographers and environmental scientists*, (London: Longman).

B. APPLICATIONS

Atwater, M.A. and J.T. Ball (1978), A numerical solar radiation model based on standard meteorological observations. *Solar Energy*, 21: 163, 170.

Ayeni, B. (1982), The testing of hypothesis on interaction data matrices. *Geographical Analysis*, 14, 79-84.

Ayeni, B. (1983), Algorithm 11: information statistics for comparing predicted and observed trip matrices. *Environment and Planning*, A 15, 1259-1266.

Baxter, M.J. and G.O. Ewing (1979), Calibration of production-constrained trip distribution models and effects of intervening opportunities. *Journal of Regional Science*, 19, 319-330.

Baxter, M.J. and G.O. Ewing (1981), Models of recreational trip distribution. *Regional Studies*, 15, 327-344.

Black, W.R. (1973), An analysis of gravity model distance exponents. *Transportation*, 2, 299-312.

Burt, J.E., J.T. Hayes, P.A. O'Rourke, W.H. Terjung and P.F. Todhunter (1980), Water: a model of water requirements for irrigation and rainfed agriculture. *Publications in Climatology*, 33, 1-119.

Chu, D.K.Y. (1982), Some analyses of recent Chinese provincial data. *Professional Geographer*, 34, 431-437.

Clark, G. and K. Ballard (1980), Modelling out-immigration from depressed regions: the significance of origin and destination characteristics. *Environment and Planning*, A 12, 799-812.

Czamanski, S. and E. Maliza (1969), Applicability and limitations in the use of national input-output tables for regional studies. *Papers RSA*, 23, 64-77.

Flowerdew, R. (1982), Fitting the lognormal gravity model to heteroscedastic data. *Geographical Analysis*, 14, 263-267.

Flowerdew, R. and M. Aitkin (1982), A method of fitting and gravity model based on the Poisson distribution. *Journal of Regional Science*, 22, 191-202.

Fotheringham, A.S. (1983), A new set of spatial interaction models: the theory of competing destinations. *Environment and Planning*, A 15, 15-36.

Fotheringham, A.S. and P.A. Williams (1983), Further discussion on the Poisson interaction model. *Geographical Analysis*, 15, 343-346.

Gaile, G.L. (1983), Reanalyses of Chinese spatial inequality. *Professional Geographer*, 35, 467-468.

Gale, N., L.J. Hubert, W.R. Tobler and R.G. Golledge (1984), Combinational procedures for the analysis of alternate models: an example from interregional migration. *Papers, Regional Science Association*, 53, 105-115.

Hathaway, P.J. (1975), Trip distribution and disaggregation. *Environment and Planning*, A 7, 71-97.

Johnson, E.R. and R.L. Bras (1978), Multivariate short-term rainfall prediction. *Water Resources Research*, 16, 173-185.

Jones, III, J.P. (1984), A spatially-varying parameter model of AFDC participation: empirical analysis using the expansion method. *Professional Geographer*, 36, 455-461.

Kim, T.J., D.E. Boyce and G.J.O. Hewings (1983), Combined input-output and commodity flow models for interregional development planning: insights from a Korean application. *Geographical Analysis*, 15, 330-342.

Lewis, P.E. (1975), An empirical test of alternative theories of trade. *Annals of Regional Science*, 9, 102-111.

Longley, P.A. (1984), Comparing discrete choice models: some housing market examples. In: *Discrete Choice Models in Regional Science*, (ed) D.E. Pitfield, (Norwich: Pion), 163-180.

Lovett, A.A., I.D. Whyte and K.A. Whyte (1985), Poisson regression analysis and migration fields: the example of the apprenticeship records of Edinburgh in the seventeenth and eighteenth centuries. *Transactions*, IBG 10, 317-332.

McMenamin, D.G. (1973), Constructing and testing a regional minimum-survey input-output table. *Institute of Government and Public Affairs Paper Series*, 182, (Los Angeles: Institute of Government and Public Affairs, UCLA).

Miller, E.J. and M.E. O'Kelly (1983), Estimating shopping destination choice models from travel diary data. *Professional Geographer*, 35, 440-448.

Nader, G.A. (1984), The rank-size model: a non-logarithmic calibration. *The Professional Geographer*, 36, 221-227.

Openshaw, S. (1976), An empirical study of some spatial interaction models. *Environment and Planning*, A 8, 23-41.

Openshaw, S. (1979), A methodology for using models for planning purposes. *Environment and Planning*, A 11, 879-896.

Openshaw, S. and C.J. Connolly (1977), Empirically derived deterrence functions for maximum performance spatial interaction models. *Environment and Planning*, A 9, 1067-1080.

Pitfield, D.E. (1978), Sub-optimality in freight distribution. *Transportation Research*, 12, 403-409.

Powell, G.L., A.J. Brazel and M.J. Pasqualetti (1984), New approach to estimating solar radiation from satellite imagery. *The Professional Geographer*, 36, 227-233.

Schaffer, W.A. and K. Chu (1969), Nonsurvey techniques for constructing regional interindustry models. *Papers, RSA* 23, 83-101.

Southworth, F. (1977), *Problems and approaches to goodness-of-fit testing for trip distribution models*, (Leeds: Alastair Dick and Associates, RHTM Project).

Southworth, F. (1983), Temporal versus other effects on spatial interaction model parameter values. *Regional Studies*, 17, 41-48.

Suckling, P.W. and J.E. Hay (1976), Modelling direct, diffuse and total solar radiation for cloudless days. *Atmosphere*, 14, 298-308.

Thomas, R.W. (1977), An interpretation of the journey-to-work on Merseyside using entropy-maximizing models. *Environment and Planning*, A 9, 817-834.

Timmerman, H.J.P. (1984), Discrete choice models versus decompositional multiattribute preference models: a comparative analysis of model performance in the context of spatial shopping-behaviour. In: *Discrete Choice Models in Regional Science*, (ed) D.E. Pitfield, (Norwich: Pion), 88-101.

Willmott, C.J. and D.E. Wicks (1980), An empirical method for the spatial interpolation of monthly precipitation within California. *Physical Geography*, 1, 59-73.

NION OF UCATIO

X